Transformation of the natural environment in Western Sørkapp Land (Spitsbergen) since the 1980s

Transformation of the natural environment in Western Sørkapp Land (Spitsbergen) since the 1980s

Edited by
Wiesław Ziaja

Authors
Wiesław Ziaja, Justyna Dudek, Maja Lisowska, Maria Olech, Krzysztof Ostafin,
Piotr Osyczka, Michał Węgrzyn

Jagiellonian University Press

Książka sfinansowana przez Ministerstwo Nauki i Szkolnictwa Wyższego oraz Uniwersytet Jagielloński
A book subsidized by the Ministry of Science and Higher Education, and the Jagiellonian University

Książka zawiera wyniki projektu badawczego MNiSW nr N N305 035634 *Zmiany środowiska przyrodniczego Zachodniego Sørkapplandu (Spitsbergen) pod wpływem globalnego ocieplenia i działalności człowieka w ostatnim 25-leciu*
The book relies on the Polish Research Project N N305 035634 *Changes in the western Sørkapp Land (Spitsbergen) natural environment due to global warming and human activity since 1982*

Autorzy

Instytut Botaniki UJ — *Maja Lisowska*, *Maria Olech*, *Piotr Osyczka*, *Michał Węgrzyn*

Instytut Geografii i Gospodarki Przestrzennej UJ — *Justyna Dudek*, *Krzysztof Ostafin*, *Wiesław Ziaja*

Reviewer — *dr hab. Piotr Gębica, prof. nadzw.*

English language editing by — *Grzegorz Zębik*

Cover designer — *Paweł Bigos*

Cover photos — *Wiesław Ziaja*

ISBN 978-83-233-3231-2

www.wuj.pl

Jagiellonian University Press
Editorial Offices: ul. Michałowskiego 9/2, 31-126 Kraków
Phone +48 12-631-18-81, +48 12-631-18-82, Phone/Fax +48 12-631-18-83
Distribution: Phone +48 12-631-01-97, Phone/Fax +48 12-631-01-98
Cell Phone 506-006-674, e-mail: sprzedaz@wuj.pl
Bank PEKAO SA, IBAN PL8 80 1240 4722 1111 0000 4856 3325

TABLE OF CONTENTS

Wiesław Ziaja

INTRODUCTION

Zdzisław Czeppe, a physical geographer, viewed Sørkapp Land as a potential geographic area for scientific research. He was a participant of the Polish expedition during the 3rd International Geophysical Year, which wintered on the northern coast of Hornsund Fjord on Isbjørnhamna bay in 1957–1958, and the summer expeditions of 1959 and 1960. Sørkapp Land, located south of the fjord, appeared to be an ideal place for research, which unfortunately could not be carried out at the time. However, he returned to western Sørkapp Land as a professor and the leader of Jagiellonian University expeditions in 1980 and 1981. He created a program of interdisciplinary research for this area, which was executed by the University's summer expeditions in the 1980s. Landscape analysis played the most important role in the research of abiotic environmental features, whereas botanical analysis was crucial in the research of biotic features. The former was carried out in 1981–1984 and 1986, and the latter in 1982 and 1985. Six physical geographers (Z. Czeppe, P. Gębica, K. Kalicki, M. Kuczek, P. Libelt and W. Ziaja) took part in field investigations of landscape, and two botanists (E. Dubiel and M. Olech) took part in field investigations of vegetation. Their published results constitute the first relatively complete and reasonably detailed (maps at a scale of 1 : 25 000–1 : 50 000) characterization of the natural environment of the area.

The natural environment and landscape in Sørkapp Land have changed rapidly since the 1990s mainly due to climate warming as well as due to the progressive regeneration of animal life since the establishment of South Spitsbergen National Park in 1973. The effects of this transformation were observed during two short (two-week) Jagiellonian University summer scientific expeditions, which covered the northernmost part of western Sørkapp Land starting with the eastern coast of Gåshamna bay: (1) at Konstantinovka hut in 2000, (2) in the tent in 2005. An understanding of this transformation, in light of the accelerated evolution of the Arctic natural environment associated with *global climate change*, could help solve new and interesting research problems such as the regeneration of the reindeer population, which was virtually absent in the 1980s.

After a quarter of a century, two participants of the aforementioned expeditions in the 1980s, Maria Olech and Wiesław Ziaja, thought of repeating these investigations using similar methods but more advanced techniques (satellite remote sensing, GPS). Hence, together with younger colleagues, they proposed a research project titled *Changes in the western Sørkapp Land natural environment due to global warming and human activity since 1982* and asked the Ministry of Science and Higher Education for financing. The project's objective was an analysis of changes in the structure and functioning of the western Sørkapp Land natural environment due to the aforementioned factors since the beginning of the 1980s. In 2008, the project was approved by the Ministry and a summer Jagiellonian University scientific expedition to western Sørkapp Land was organized.

The research team of this expedition was comprised of six people: Wiesław Ziaja (geographer, leader), for whom it was the ninth summer research season in Sørkapp Land and twelfth in Spitsbergen, Piotr Osyczka (botanist), for whom it was the second summer season in Spitsbergen, Michał Węgrzyn (botanist), Justyna Dudek (geographer), Maja Lisowska (botanist), and Jan Niedźwiecki (geographer), who were spending their first summer in the Arctic. Five expedition members arrived in Longyearbyen by air on June 30th, 2008. Piotr Osyczka arrived via a research vessel operated by Gdynia Maritime University, *Horyzont II*. He arrived at the Polish Polar Station (of the Institute of Geophysics, Polish Academy of Sciences) in Hornsund with the main expedition's baggage. After their arrival at Longyearbyen, the expedition members found shelter at Marek Zajączkowski's hospitable home. From Longyearbyen, a yacht called the *Eltanin*, under Jerzy Różański's (owner and shipmaster) command, transported them to the westernmost part of the Sørkapp Land coast. The first attempt (July 1–3) failed because of a dense ice-pack at the mouth of the Hornsund Fjord, and the yacht had to return to Longyearbyen. After a few days, three members of the expedition arrived from Longyearbyen to Hornsund via the *Eltanin* and the remaining members arrived via the *Horyzont II* on July 8th. The three expedition members who had arrived at the Polish Polar Station first, had boarded the yacht during its brief stop in Isbjørnhamna bay, which is near the station. Immediately, the yacht sailed across the mouth of Hornsund Fjord to its southern shore in front of the trapper hut north of Palffyodden. Our colleagues from Hornsund Station, under Mateusz Moskalik's command, accompanied the yachts in motorboats loaded with the expedition's baggage (food, built materials, equipment). The expedition was conducted from the yacht to the Sørkapp Land coast with their help, landing on the coast on July 8th at 6:00 p.m. The staff of the Polish Polar Station in Hornsund, under the leadership of Piotr Głowacki and Marek Szymocha, helped our expedition in a number of ways, transporting our baggage and delivering necessary equipment several times. The geological team from the Polish Academy of Sciences in Warsaw, under the command of Krzysztof Krajewski, transported us via their dinghy to Breinesflya and Gåshamnøyra. Jerzy Czerny, our colleague from the AGH University of Sciences and Technology in Cracow, took one of the expedition members (who had to go away earlier from Spitsbergen) by boat from Sørkapp Land. On August 12th at 3:00 a.m., the expedition was transported by Krzysztof Krajewski from the coast near Palffyodden to the *Eltanin* yacht and departed from Sørkapp Land. We are very obliged to the aforementioned colleagues and their teams for their priceless help, which enabled us to realize all the elements of our plan. The expedition returned to Poland on August 15th.

Wiesław Ziaja, Krzysztof Ostafin

STUDY AREA

Why is Sørkapp Land, and specifically western Sørkapp Land, such an attractive area for scientific research?

The answer is: Due to its geographic location and natural environmental features that make it unique in this part of the Arctic.

Sørkapp Land is the southern peninsula of Spitsbergen, the largest island of the Svalbard Archipelago. The island's east-west width is 150–200 km in its northern part. The island narrows to the south in the shape of a wedge between the Barents Sea in the east and the Greenland Sea in the west (Fig. 1). Their waters differ in temperature, which affects local climate conditions. The eastern coast is affected by the cold East Spitsbergen Current, which flows from the Arctic interior towards the south. This results in extensive glaciation and the lack of continuous vegetation. On the other hand, the western coast is warmed from the south by the West Spitsbergen Current, the last branch of the Gulf Stream. This produces relatively little glaciation and allows for a continuous tundra. However, southern winds cause a narrow belt of the cold current's waters to separate the warm current from the coast. Moreover, the two coasts differ in terms of geological structure and relief. In the east, the mountains are built of non-resistant Cretaceous and Tertiary rocks that fall down directly into the sea or onto narrow coastal lowlands (less than 1 km wide) at some locations. In the west, the mountains are separated from the sea by much wider (up to 4 km) coastal lowlands, with both cutting into hard Proterozoic and Paleozoic rocks (see cover photo). For this reason, the two Spitsbergen coasts are extremely different in terms of their natural environment and landscape (Hisdal 1985; Stange 2003). The further to the south, the smaller the distance between the two coasts.

This is true primarily of Sørkapp Land because the peninsula constitutes the end of the land wedge. Its maximum east-west width is 40 km in the north, and only 17 km in the south. Moreover, this wedge is "torn" in two places. In the north, two fjords, Hambergbukta from the east and Hornsund from the west, leave only a narrow isthmus between them. The width of the isthmus decreased to a mere 7 km in 2011 as a result of the present-day retreat of tidewater glacier fronts at the fjord heads due to the warming of the climate. The peninsula gradually narrows also in the south due to the retreat of the ice cliffs of the Vasilievbreen (east) and Olsokbreen (southwest) glaciers. The distance between the glaciers' cliffs has decreased to 12 km. The north-south length of the peninsula is 55 km (Fig. 2).

Fig. 1. Satellite image of southern Spitsbergen (south of Van Mijenfjorden) from August 20th, 1985: Landsat 5, MSS-421. Color scheme: white – glaciers, red or reddish – vegetation cover, beige and blue-grey – unglaciated areas without vegetation cover, black – seawater, blue-and-black – glacial water mixed with seawater

►

Fig. 2. Satellite image of Sørkapp Land consisting of four *TerraASTER* scenes from 2003–2005. Color scheme: blue with white – glaciers, red or reddish – vegetation cover, grey-greenish – land areas without vegetation cover, blue stripes or dots against a dark background – ice-pack on the sea, dark with greenish stripes – seawater with admixtures of glacial water. Majority of the image is current for 2004

All of these factors help generate a great deal of variety in the peninsula's natural environment: from the western and southern lowlands overgrown by tundra with herds of reindeer to the glacial-mountainous Arctic desert in the interior and east. Therefore, the peninsula is an excellent study area featuring all the relationships between the different components of the natural environment. They result in an unusually diverse, completely natural and almost primeval landscape: glacial and periglacial, mountainous and low-lying, inland and coastal. These basic landscape types are internally differentiated: e.g. fjord-type coastal landscape and open-ocean-type coastal landscape. Traces of the Pleistocene ice sheet may be discovered in the contemporary landscape. In addition, the reaction of this environment to climate warming can be readily noted due to relatively rapid climate fluctuations since the 1980s. New climate phenomena and associated trends can be easily observed in this area.

Apart from that, Sørkapp Land belongs to the least accessible by sea Arctic areas, mainly because of wide shallows off-shore, numerous submerged rocks and rela-

Hornbreen
Hambergbr.
Hambergbukta
Ostrogradskifjella
Isbjørnhamna
Hornsund
Körberbr.
Samarinvågen
Gåsh.
Wurmbrandegga
Hornsundtind
Daudbjørn-
pynten
Kulmstranda
Lisbetdalen
Gasbreen
Hornsundneset
Struvefjella
Gavrilov-
fjellet
Bungebreen
Vasilievbreen
Isbukta
Breinesflya
Vitkovskibreen
Tørrflya
Hilmar-
fjellet
Olsokbreen
Olsokflya
Stormbukta
Øyr-
landet
Kistefjellet
Sørflya
Sørkappøya
10 km

tively new bays and fjords at the tidewater glaciers' retreating fronts, which have not been surveyed yet.

Western Sørkapp Land (Fig. 3) is the most interesting part of the peninsula.

The area is accessible primarily by sea, as it protrudes between the open Greenland Sea and Hornsund Fjord, which results in the shortest season with sea-ice, i.e. the longest period without a barrier to newcomers. Moreover, there is only one Sørkapp Land anchorage in Gåshamna bay near the fjord's southern coast, close to its mouth.

Sørkapp Land was under the Barents Ice Sheet at least once during the Pleistocene, most likely during the Late Weichselian (Salvigsen, Elgersma 1993). The western part of the peninsula was completely covered by the ice sheet, which is evidenced by erratics on Kovalevskifjellet peak at the elevation of 640 m (Ziaja 1989).

Most of western Sørkapp Land has been devoid of glaciers (Ziaja 1999) throughout the Holocene. The unglaciated area includes not only the coastal lowlands of Kulmstranda and Hornsundneset but also two mountain ranges: Struvefjella and a second unnamed range, which consists of Wurmbrandegga, Savitsjtoppen, Kovalevskifjellet and Gavrilovfjellet. The unglaciated area also includes Lisbetdalen valley between the two mountain ranges, which is an exception in Sørkapp Land and a rarity in Spitsbergen. Animals, plants and soils are very common and fully visible in the area's landscape. This is not true in most parts of the peninsula. Little research had been conducted in the area until the 1980s. However, some hunting activity had been conducted since the 17th century.

Animal life, plant life and soils are also well developed on the low (elevation: below 50 m above sea level) coastal plains in the southern part of western Sørkapp Land, which is covered in part by tongues of glaciers extending from the peninsula's interior. Adjoining mountains are also partly glaciated on the land side. Two coastal plains, the wide Breinesflya plain at the foot of Wiederfjellet massif and the narrow Tørrflya plain at the fore-field of Bungebreen glacier, extend from the northwest to the southeast. Furthermore, narrow karst plains stretch at the base of the mountains from the Bungeeleva pro-glacial river to the lateral marginal zone of the large Olsokbreen glacier.

The northernmost part of western Sørkapp Land consists of a low plain along Gåshamna bay and Tsjebysjovfjellet mountain, with a narrow lowland at its base on the fjord to Körberbreen glacier in the east. This area, together with Gåsbreen glacier, was already investigated in detail in 1899 (De Geer 1923). Hence, the monitoring of changes in Gåsbreen glacier is the longest of all more than 80 of the Sørkapp Land glaciers. Gåsbreen glacier is a part of the boundary between the high-mountain-glacial interior and western part of the peninsula.

The lack of glaciers and a compact area of about 100 km^2 make this area relatively safe for exploration, which was particularly important for the early expeditions in the 1980s.

Animals in Sørkapp Land and the adjoining seas have been hunted since the 17th century. Historical remains including graves of Western European whalers and

Fig. 3. Satellite image of western Sørkapp Land consisting of *TerraASTER* scenes. Color scheme: light-blue with white – glaciers, red or reddish – vegetation cover, brown – land areas without vegetation cover, dark-blue with lighter dots – seawater with ice-pack. The image is current for 2004

Pomor (North-Russian) hunters have been preserved in the western part of the peninsula.

The last Norwegian trapper station was operational until the establishment of South Spitsbergen National Park (south of Van Keulenfjord) in 1973 and has been preserved on the coast, one kilometer north of Palffyodden cape in an area most abundant in game. The station is a good base for summer scientific expeditions because its main hut is built of timber beams (two cozy rooms with stoves and a porch), making it resistant to polar bear attacks (see upper back cover photo). A very small auxiliary hut (ca. 7 m^2, 1.5–1.7 m high), situated on the Breinesflya coast at a distance of about a half a day by foot from Palffyodden, also belongs to the former trapper station. The smaller hut, also built using timber beams, is a safe shelter for two individuals (see lower back cover photo).

The three advantages of western Sørkapp Land – relatively easy access, lack of glaciers and a former trapper station – are the more valuable due to the presence of many interesting environmental phenomena and historical sites.

In the 1980s, researchers went on foot from the huts to other parts of western Sørkapp Land. In most cases, they did not cross Körberbreen glacier in the north and Olsokbreen glacier in the south, as the glaciers flow towards the sea.

The glaciated peninsula's interior is extremely different from western Sørkapp Land because biotic features are rare and hardly visible in the landscape of the peninsula interior. The boundary between the two regions is shifting slowly to the east due to the retreat of the glaciers.

METHODS AND MATERIALS

Wiesław Ziaja, Justyna Dudek

LANDSCAPE RESEARCH

Complex landscape field mapping at a scale of 1 : 25 000 was the basic method of surveying the study area in the 1980s. A total of 1,514 small basic landscape units (so-called geocomplexes of the *uroczysko* range) were mapped across an area of ca. 94 km^2. An old Norwegian topographic map at a scale of 1 : 100 000 (Sørkapp sheet), current for 1936 and enlarged fourfold, was used as a base map. Each individual landscape unit was drawn on the map, numbered and described on a special separate form. The demarcation criteria consisted of spatial changes in environmental and landscape features. The following characteristics were analyzed: elevation, slope gradient and exposure, lithostratigraphy, tectonics (dip of the rock strata), Quaternary deposits (genesis, granulation, thickness), morphogenesis, micro-relief, geomorphic processes, meso-climate (based on exposure, shadowing and persistence of snow patches), bodies of water and vegetation (density and composition, division into vascular plants, mosses and lichens). The mapping covered virtually all of the area that had not been covered by glaciers during the Holocene in western Sørkapp Land. The landscape units were grouped into several dozen types, which made it possible to create detailed maps of the landscape structure of the area (Czeppe, Ziaja 1985; Kuczek, Ziaja 1990; Ziaja 1991, 1992). This structure did not change by 2008. However, certain changes within some of the landscape types were observed in 2008.

In addition, the extent of glaciers and their marginal zones as well as the sea coastline were mapped via a traditional method (using a *Paulin* altimeter) in the 1980s, and using more modern methods (GPS *12 Garmin*) in 2008.

The interpretation of aerial photographs obtained from the Norsk Polarinstitutt, especially black-and-white photographs from 1961 and infrared photographs from 1990, as well as the analysis of satellite data were combined with the results of field investigations in a very effective way.

The analysis of glacier retreat since the 1980s was divided into two parts. The first part addressed changes in the glaciers' surface area in 1983–1984, 1990, 2003, 2004, 2007 and 2008 (Maps 1 and 2). In the second part, changes in the glaciers' surface elevation from 1990 to 2008 were inferred from two digital elevation models (Maps 3 and 4).

The following data sets were used in the manual delineation of the extent of each glacier:

1. Three sheets of a topographic map (Hornsund, Gåsbreen, and Bungebreen) at a scale of 1 : 25 000, edited by the Polish Academy of Sciences in 1987. The map was compiled photogrammetrically on the basis of vertical aerial photographs at a scale of 1 : 50 000 (taken in summer 1961) and geodetic measurements carried out during the 6th Spitsbergen Expedition of the Academy during the 1983 and 1984 summer seasons.
2. A topographic map sheet (C13 Sørkapp) at a scale of 1 : 100 000, edited by the Norwegian Polar Institute in 2007. The map illustrates the extent of each glacier and provides contours at 50 m intervals. The part of the map with the study area was created based on vertical aerial photographs at a scale of 1 : 50 000 (taken in summer 1990).
3. ASTER data with 15 m spatial resolution provided by NASA's Earth Observing System (EOS), and distributed by the Land Processes Distributed Active Archive Center (LPDAAC): one scene acquired on July 24th, 2003, two consecutive scenes acquired on August 7th, 2004, and two consecutive scenes acquired on August 14th, 2007. All ASTER bands used in the study were orthorectified and projected to WGS84, UTM zone 33 north projection.
4. SPOT 5 Orthophoto with 5 m spatial resolution, acquired on September 1st, 2008. The data were provided by the IPY SPIRIT (International Polar Year SPOT5 stereoscopic survey of Polar Ice: Reference Images and Topographies) project (Korona et al. 2009). The mapping of glaciers was difficult in 2008 due to thick snow cover; therefore, only the frontal sections of tidewater glaciers and Bungebreen were updated using this data set.

The maps were scanned and georeferenced to a common datum and projection (WGS84, UTM zone 33 north) using the four corners and 50 evenly distributed points for each map sheet edited by the Polish Academy of Sciences and about 100 points for the map edited by the Norwegian Polar Institute. The extent of each glacier based on Polish maps was validated and corrected using photographs taken in the field during Jagiellonian University expeditions in 1983 and 1984.

Changes in glacier surface elevation were estimated using two digital elevation models (DEMs). In the first DEM, contours were interpolated at 50 m intervals based on the new Norwegian topographic map (C13 Sørkapp) at a scale of 1 : 100 000. The DEM was generated using ESRI®ArcMap™ 9.3.1 software via an interactive finite difference technique (Hutchinson 1989) and has a grid spacing of 40 m. The first DEM was used as a reference data set. The second DEM, with 40 m pixel size, was provided by the IPY SPIRIT project. It was generated from a cloud-free high resolution SPOT 5 HRS image pair acquired on September 1st, 2008. The DEM extraction was performed by the French Institut Géographique National (IGN) following an automatic processing method, including manual intervention and interactive checks against ground-based measurements (Korona et al. 2009). The optical contrast of the images was mostly satisfactory, although in some parts, image matching had failed. Unmatched areas were masked and marked as "error in data." The SPIRIT DEM was co-registered using an analytical solution developed by Nuth and Kääb (2011), based on the cosinusoidal relationship between elevation differences and elevation derivatives of the slope and aspect. The method permitted to minimize a three-dimensional shift vector between two DEMs.

Changes in glacier elevation over the period 1990–2008 were estimated simply by subtracting the SPIRIT DEM from the reference DEM. The resulting values in the new raster indicated the difference in elevation at each pixel.

The work was performed in part at the Department of Geosciences, University of Oslo, with support granted by Iceland, Lichtenstein and Norway by means of co-financing from the European Economic Area Financial Mechanism and the Norwegian Financial Mechanism as part of the Scholarship and Training Fund.

Michał Węgrzyn, Maja Lisowska, Maria Olech, Piotr Osyczka

BOTANICAL RESEARCH

During the 1982 and 1985 summer seasons, extensive vegetation research was carried out in northwest Sørkapp Land. The study area included marine terraces along the northern and western shores of Sørkapp Land, from the Lisbetelva river to the Vinda river, Hohenlohefjellet (614 m), as well as the western slopes of Sergeijevfjellet and Lidfjellet.

Western Sørkapp Land is diverse in terms of habitat conditions such as: topography, bedrock, hydrological and edaphic relationships and microclimate. Lichens and bryophytes are predominant along with a few flowering plants, creating a complex mosaic of vegetation (Fig. 4).

Phytosociological research led to the identification of 28 vegetation units in the study area. The basis for the identification was 285 phytosociological relevés, performed according to the Braun-Blanquet (1964) method. Relevés were taken at different locations in order to obtain a full picture of the variety of vegetation. Complete phytosociological tables were created for 28 plant communities (Dubiel, Olech 1990). Plant communities were selected on the basis of their floristic characteristics. In some cases, the presence of dominant species strongly influencing the physiognomy of patches was considered more important. A short description of the habitat as well as comments on the distribution were added to the descriptions of vegetation. Names of differential species and their ecological scale were given for each community. Community names derive from the names of differential species, which were often dominant species. No new names of associations were created, as that would be premature, given the current state of knowledge of the diversity of vegetation in Spitsbergen.

Next, field mapping was carried out. The arrangement of plant communities identified during phytosociological research was marked on a 1 : 25 000 map (Dubiel, Olech 1991). Vegetation units were marked on the map for four primary categories: a) homogeneous plant communities occupying large areas; b) examples of clear dominance of one community over others within a particular mosaic; c) arrangement of two or three neighboring communities creating a mosaic; d) large areas of transition between two or more communities.

In 2008, phytosociological research was repeated in the same plant communities, which had been identified in the 1980s (Dubiel, Olech 1990). The Braun-Blanquet method (1964) was used once again. In phytosociological relevés, all groups

Fig. 4. Central part of the Hornsundneset terraced coastal plain built of Early Carboniferous sandstones, siltstones and shales. A general view of the tundra vegetation and lakes (dammed by isostaticly raised coastal ridges) from the slopes of Sergeijefjellet, facing WNW. Photo: M. Węgrzyn, 2008

of plants and fungi contributing to the composition of communities were included. The repetition of the phytosociological study in the same geographic area provided an accurate record of the state of the vegetation, and also allowed to evaluate the speed and direction of changes. As was the case with earlier research (Dubiel, Olech 1990), the exact syntaxonomic position of certain plant communities was not precisely specified, since it was not the aim of the research. However, the synonyms of the names of plants and fungi were updated.

In order to obtain data about changes in the size of plant communities, vegetation mapping was repeated on a 1 : 25 000 scale map. Field GPS devices containing a rectified version of the vegetation map from the 1980s, with areas of different vegetation units marked, were employed during the fieldwork. In many cases, the location of certain hydro-geomorphological elements such as streams, rivers, lakes, slopes and shorelines, all of which create natural boundaries for plant communities, was adjusted as well.

Both vegetation maps were substantially improved. Relief was added to the maps, which created an opportunity for easier interpretation of the extent of each plant community. The final versions of both maps were prepared on the same scale of 1 : 50 000 in order to show vegetation changes more accurately (Maps 5 and 6).

Vegetation maps of the same geographic area created for different periods of time are the best way to show spatial changes in vegetation.

COMPONENTS OF NATURAL ENVIRONMENT

Wiesław Ziaja

BEDROCK

Solid pre-Quaternary bedrock in western Sørkapp Land can be grouped into three very different complexes:

1) the most extensive and oldest rocks, from the Middle Proterozoic to the Silurian, folded in the Caledonian Orogeny, formerly called Hecla Hoek, and Pre-Old Red basement today. This includes different types of dolomites, phyllites, schists, quartzites, limestones, sandstones, breccias and others; between them, Pre-Cambrian rocks are rather high-grade methamorphic rocks and Palaeozoic rocks are mainly low-grade methamorphic rocks;
2) Early Carboniferous clastic sediments in the northwest of the peninsula (protruding between the open Greenland Sea and Hornsund Fjord): sandstones, including quartzitic sandstones, siltstones and shales (see cover photo);
3) Triassic sedimentary rocks that occur discordantly at some locations on the two aforementioned complexes: mainly sandstones and conglomerates.

None of these complexes lies horizontally. The rocks of the Pre-Old Red basement feature the largest dips: up to 90° and more (overturned folds). Dips of the next two complexes' rock strata tend to be much smaller: from 10 do 25°, sometimes more. All the complexes are cut by faults. The most important fault direction is NNW–SSE (Winsnes et al. 1992; Dallmann et al. 1993; Dallmann 1999).

Loose (unconsolidated) Quaternary deposits (marine, glacial, glacifluvial, fluvial, organic, frost-weathering, eolian, slope, including talus and solifluction) are not continuous.

Coastal plains are partly covered by thin (up to 2–3 m), fragmented in many places, deposits of the Pleistocene and Holocene marine accumulation (rounded boulders and gravels, and sand). They are weathered to a substantial degree, apart from boulders and gravels of the Early Carboniferous quartzitic sandstone. Tørrflya, built of marine deposits up to ca. 10 m thick, is the exception.

Glacial deposits, both moraines and glacifluvial deposits, may be thicker in places (up to several meters), especially on coastal plains and mountain valley floors. Most are very young Holocene – from the Little Ice Age and the 20th–21st century (i.e. from the past several centuries), rarely from the end of the Pleistocene, such as numerous limestone (of the Slaklidalen formation) erratics.

Steep mountain slopes cut in solid rock are devoid of any superficial Quaternary deposits due to erosion-denudation processes which remove any regolith down.

Talus and talus-torrent fans up to 10–20 m thick, in places connected into a one talus slope, lie at the bottoms of slopes. Less steep slopes (under 28–30°) can be overlain by weathering-solifluction deposits 1–2 m thick; the smaller the slope gradient, the more frequent and the thicker the deposits.

Flat areas (0–3°) are mostly overlain by weathering (on top plateaus) or accumulation types of cover, which can be as much as 2 m thick. Flat areas undergo further weathering and frost segregation. A thin (up to 30 cm) soil-vegetation cover is deposited across flat areas.

A unique superficial micro-relief is a feature of each type of Quaternary deposit.

Wiesław Ziaja

CLIMATE

The climate in Sørkapp Land and throughout Spitsbergen is Arctic and marine-type. Average annual temperatures vary from –9°C to –3°C in the west of Sørkapp Land, i.e. the temperatures are quite high in relation to the latitude of almost 77°N, due to the warm West Spitsbergen Current. Average summer (July and August) temperatures are usually in the range of 3–5°C, thus they are relatively low due to the ocean's influence. Average annual precipitation totals are usually in the range of 300–500 mm. Average summer (July and August) precipitation totals are much more variable and range from 30 to 200 mm in different years. Since the beginning of the 20th century, no yearly meteorological observations have been carried out in western Sørkapp Land. Hence, the aforementioned climate characteristics were based on meteorological observations carried out at the station on Isbjørnhamna bay, which belongs to the Institute of Geophysics of the Polish Academy of Sciences (Marsz, Styszyńska 2007). The station is located north of Hornsund Fjord at a 10 km distance from the northern coast of western Sørkapp Land. The data from this station are compatible with the results of meteorological observations carried out at different locations in western Sørkapp Land during selected summer seasons in the 20th century (Ziaja 1999).

Climate fluctuations during the 20th century are very important in the context of this book's subject matter, especially the last fluctuation, which has occurred since the 1980s and is described below in the chapter on environmental changes.

The climate in western Sørkapp Land is unique in terms of the entire peninsula and southern Spitsbergen because its features indicate a lack of glaciers in the local mountains throughout the Holocene. These mountains are not lower than mountains in the rest of the peninsula, which are more or less covered by glaciers. The mountains in northwest Sørkapp Land remain free of glaciers due to their exposure to relatively warm and dry eastern foehn winds, which melt the winter snow cover. This occurs not only today but has been occurring since the beginning of the Holocene, even during periods of climate cooling. The lack of glaciers in the Holocene has been shown by geological (Salvigsen, Elgersma 1993) and geomorphological (Ziaja 1989) investigations. The other course of Late Quaternary evolution of the area was outlined by Lindner and Marks (1993). The foehn effect was detected on the basis of meteorological observations (Kalicki 1985; Ziaja 1985).

Wiesław Ziaja

GLACIATION

Sørkapp Land glaciation is clearly Artic-type for the following two reasons:

1) common presence of permafrost, which is predominantly frozen solid rock and only in the Quaternary deposits consists of ice mixed with rock material of variable granulation,
2) very weak influence of altitude (above sea level) on the distribution and extent of glaciers.

The northern part of western Sørkapp Land (Fig. 3) includes ca. 50 km^2 of mountains (seven massifs arranged into two ridges with Lisbetdalen valley between them) without glaciers during all of the Holocene, in spite of the fact that their quite extensive and flattened peaks reach an elevation of 640 m (Ziaja 1992, 1999). This is due to the exceptional local climate described earlier (equally high or lower areas situated a few kilometers further to the east and southeast have been superficially glaciated to a substantial degree).

These areas with expansive glaciers flowing from the glaciated peninsula's interior to the west and south, and with small glaciers on coastal mountain massifs, are not part of western Sørkapp Land. The extent of each glacier marks the natural boundary of the western region. A retreat of this extent is simply a shift of this boundary to the east.

The coastal lowlands – which stretch to the west of Körberbreen glacier on Hornsund Fjord and further to the southeast to Olsokbreen glacier on the open Greenland Sea – were also unglaciated throughout the Holocene. The unglaciated slopes of mountains falling to the lowlands of the peninsula's middle west (Breinesflya and others to the southeast) may also be thought of as western Sørkapp Land. However, these mountains are more or less glaciated from the land (interior) side.

Thermo-mineral springs indicate the existence of a talik, which is not covered by a glacier or lake in the southernmost corner of western Sørkapp Land (in Bjørnbeinflya and Olsokflya), analogically to what has been described by Salvigsen and Elgersma (1985).

Wiesław Ziaja

TERRAIN RELIEF

The principal landform types in western Sørkapp Land are coastal lowlands, mountains and mountain valleys (Fig. 3).

Coastal lowlands constitute more than half of the study area and were formed as a shallow sea bed offshore during the Pleistocene and the Holocene, and uplifted due to isostatic movements afterwards.

There are two coastal lowlands along Hornsund Fjord, which are isolated from other lowlands by the slopes of the Tsjebysjovfjellet and Wurmbrandegga massifs running down to the sea in the northeastern part of the study area. The narrow (up to 250 m wide) and short (ca. 1.5 km long) lowland with the Stonehengesteinane group of rocks lies at the foot of the northern wall of Tsjebysjovfjellet called Rasstupet. The second lowland, Gåshamnøyra, consists of the virtually flat outlet of a wide valley covered by extensive (2 km x 2 km) extramarginal sandur formed by proglacial waters of Gåsbreen glacier.

The Kulmstranda lowland (1.5–2.0 km wide), built of resistant Early Carboniferous sandstones, adjoins from the south of the outlet of Hornsund Fjord. The lowland is formed of extensive rock terraces, which are covered by a thin layer of quartzitic marine pebbles or devoid of them in many places. The lowland's higher part forms a ridge reaching 123 m. Lisbetdalen valley is located behind this ridge. Kulmstranda is incised by the gorge of the Lisbetelva river (ca. 10 m deep). The second, more shallow and already inactive, incision in the lowland is situated 300 m further to the east. This lowland experiences extremely strong and frequent winds, which blow away fine-grained rock material and limit plant life.

The Hornsundneset lowland, 1.5–3.0 km wide, with the hut near Palffyodden, stretches from the fjord's outlet towards the south and southeast. Successive (ever older) marine terraces are formed in tiers from the coastline up to the foot of the Struvefjella mountain range, at an altitude of 5–70 m (Fig. 4, cover photo). As was the case in Kulmstranda, the terraces also cut into resistant Early Carboniferous quartzitic sandstones. The terraces' small dip into the land side brings out their edges, and their rock surfaces are mostly covered by old coastal ridges built of marine pebbles. The lowland is fertilized by small auks *Alle alle* that nest in the mountain slopes above it, which helps promote vegetation and soil formation.

Breinesflya, the largest of the Sørkapp Land lowlands, is completely different from the aforementioned lowlands: 3–4 km wide, boggy, gradually rising from the beach to the foot of Wiederfjellet massif at an elevation of ca. 50 m, built of less resistant Triassic sedimentary rocks. The lowland is almost completely covered by a thick (most often 1–2 m) layer of fine-grained marine sediments along with an admixture of material washed away from the slopes above them as well as fluvial material ranging from clay to debris. This type of Quaternary cover favors vegetation and soil formation (Fig. 5).

Three narrow (up to 1 km wide) lowlands lie further to the southeast. Tørrflya is the accumulation marine terrace up to 10 m high and thick, with the cliff coast on Triassic bedrock, between the formerly major Vinda river and the large Bungeelva glacial river. The extra-marginal sandurs of Bungebreen glacier adjoin Tørrflya. The next lowland, Bjørnbeinflya, stretches to the southeast of the Bungeelva river. This lowland is at first more narrow between Vitkovskibreen glacier's marginal zone and the sea, and then wider at the foot of Hilmarfjellet, featuring erosion karst forms such as holes and underground channels. The next lowland, Olsokflya, is located along Stormbukta bay. Thin Triassic layers overlie the limestone Pre-Old Red basement only along the coastal belt up to 300 m wide.

The large (5 km x 4 km) Stormbukta bay did not exist at the end of the Little Ice Age in 1900 (Wassiliew 1925). The only tidewater glacier found along the western

Fig. 5. Northwestern Sørkapp Land seen from the Breinesflya low and wet coastal plain to the northwest. The following mountains are visible in the background (from left to right): Hohenlohefjellet and Sergeijevfjellet (joined together), Lidfjellet, Gavrilovfjellet, and the northwestern fragment of Wiederfjellet. Photo: J. Niedźwiecki, 2008

Sørkapp Land coast, Olsokbreen, covered the bay's present-day area. At the time, the glacier was much thicker and larger than it is today. Its northern (right) lateral moraine has changed into a wide marginal zone due to the glacier's retreat. A new coastline has also appeared in the area.

All the coastal lowlands are shaped by weathering and frost segregation, which can be observed in the form of characteristic micro-relief.

Mountains dominate the landscape despite the fact that they occupy less than half of western Sørkapp Land. The largest quantities of water and deposits flow to the lowlands and the sea from the mountains due to higher precipitation and intensive erosion-denudation processes taking place therein.

Struvefjella, the westernmost mountain range, consists of three different massifs (Fig. 5). Hohenlohefjellet (Fig. 6), the northernmost and the highest (> 600 m) of the massifs, cuts into resistant Lower Carboniferous sandstones. It is primarily covered by quartzitic debris, which creeps gravitationally (especially on the eastern slopes) or reaches its angle of repose (especially on other slopes). High nivation moraines, which often overlie old glacial moraines, and talus and talus-torrent fans are situated at the foot of the slope and slope gullies. High nivation moraines are built of coarse sandstone blocks. Sergeijevfjellet is a ridge 3 km long and built of less-resistant Lower Carboniferous and Triassic sedimentary rocks, thus lower (ca. 400 m) and with more gentle slopes (up to 30°) under a weathering-solifluction cover. The high (> 500 m) Lidfjellet massif is shaped like a truncated pyramid built of Mesozoic sedimentary rocks. Its steep slopes primarily undergo weathering and gravitational creeping. Its top is shaped like a triangle with a regular inclination gradient of ca. 10°.

The Struvefjella range is dissected by two narrow, short (1.0–1.5 km) and deep (altitude: below 145 m) tectonic valleys: Hohenloheskardet (Fig. 6) and Sergeijevskardet, shaped by the Barents Ice Sheet and the sea during the Pleistocene as well as slope processes during the Holocene. There is a Young Pleistocene glacial moraine ridge at the mouth of the Sergeijevskardet. There are also numerous limestone (Slaklidalen formation) erratics in the valleys and on the slopes of Sergeijevfjellet and Lidfjellet.

The Lisbetdalen mountain valley (Fig. 6), located east of Struvefjella, is expansive and cuts deep (altitude: below 180 m) into resistant Lower Carboniferous sedimentary rock. Traces of the presence of the Pleistocene ice sheet (deep basin of Svartvatnet lake, limestone erratics) and the sea (marine gravels) can be observed in this area. The valley was shaped during the Holocene by weathering, solifluction, sheet wash, nivation and fluvial processes. The lateral Kovalevskidalen falls into the Lisbetdalen from the east.

The second unglaciated (and unnamed) mountain range in western Sørkapp Land towers over the Lisbetdalen from the east. The range consists of the following four mountain massifs: the narrow and steep Wurmbrandegga ridge built of Proterozoic dolomites (ca. 400 m), and further to the south, the Savitsjtoppen (almost 500 m), the Kovalevskifjellet (640 m) and the Gavrilovfjellet (almost 600 m) mas-

Fig. 6. Lisbetdalen valley with Svartvatnet lake seen from the east. In the background: the eastern slopes of Hohenlohefjellet, which fall down to the lake and the northeastern slopes of Sergeijevfjellet. Large talus-torrent fans and nival moraines are visible at the base of Hohenlohefjellet on the lake. Photo: W. Ziaja, 2008

sifs. More gentle western slopes and top plateaus of the last three massifs are primarily built of Lower Carboniferous and Triassic sedimentary rocks and have been shaped by geomorphic processes similar to those in Lisbetdalen (apart from fluvial processes). The steep eastern slopes of these three massifs, built of non-resistant Proterozoic phyllites, undergo intensive weathering and creeping. These slopes fall down to two valleys, which form the boundary between western and central Sørkapp Land: the northern unnamed valley with Gåsbreen glacier's marginal zone and the Slaklidalen valley, which runs to the south.

Limestone erratics can be found on the top plateau of Kovalevskifjellet. Hence, this mountain range must have been covered by the ice sheet at the end of the Pleistocene (Ziaja 1989; Salvigsen, Elgersma 1993).

The deep and narrow V-shaped Liddalen valley can be found at the junction of the two aforementioned ranges (Sruvefjella and the unnamed range), south of Lisbetdalen, between Gavrilovfjellet and Lidfjellet.

The long (6 km) and high (> 700 m) Wiederfjellet-Stupryggen mountain range stretches to the southeast from Slaklidalen to Bungebreen glacier. This range, wide in the northwest and narrow in the southeast, is built of different types of rocks of the Pre-Old Red Basement. The slopes of the range's northwestern part are cut in the aforementioned phyllites and undergo weathering, sheet wash and solifluction. The rest of the range, built of dolomitic sandstones and metamorphic limestones, is extremely asymmetric. Its gentle eastern slopes are covered partly by Wiederfjellet glacier. Its steep western slopes (weathering and creeping) fall down to the bend (on the fault) between the range and the Breinesflya coastal plain. A narrow belt of gentle solifluction slopes has formed on this bend.

The Plogen massif, almost 700 m high and with steep slopes, is located between the valleys of the Bungebreen and Vitkovskibreen glaciers. The Hilmarfjellet-Karentoppen massif (> 800 m) is located in the southeastern corner of the study area between the Vitkovskibreen and Olsokbreen glaciers. Both massifs are built of metamorphic limestones and sandstones of the Pre-Old Red Basement, while their peaks are built of Triassic sedimentary rocks. The highest marine raised terraces in Spitsbergen (up to 338 m) are preserved on the slopes of Hilmarfjellet (Werenskiöld 1953).

The last three mountain massifs are at least partly glaciated from the land side. However, their slopes, which fall down to the coastal lowlands, are free of glaciers.

Wiesław Ziaja

WATERS

Karst springs can be found at the foot of Tsjebysjovfjellet on Hornsund Fjord (Pulina 1977; Leszkiewicz 1982) in the northeastern corner of the study area. Gåshamnøyra plain, a large extramarginal sandur located west of the massif, was formed by streams from both Gåsbreen glacier and the valley south of the glacier's front edge. These streams flow across the glacier's marginal zone built mainly of dead ice undergoing rapid melting.

Non-glacial rivers and lakes play an important part in the western Sørkapp Land water system. This is not true elsewhere in Spitsbergen. The rivers and lakes in western Sørkapp Land are supplied directly by atmospheric precipitation – including snow patches, which thaw before the following winter – and an active layer of permafrost. This is especially true of the northern part of the study area due to a complete lack of glaciers there. The largest body of fresh and clean water is Svartvatnet lake (Fig. 6), which is found in a deep glacial erosion trough. The lake's water table is found at an altitude of 72 m above sea level. The lake contains a population of *Salvelinus alpinus* L. salmonoid fish (Gullestad, Klemetsen 1997). Two streams flow into the lake: (1) a main stream, which flows from the snow patch located in the pass (at almost 200 m) between the Lidfjellet and Kovalevskifjellet massifs, and (2) a lateral stream flowing from Kovalevskidalen valley. The entire Lisbetdalen valley is drained by the Listbetelva, which is the only permanent river in the northern part of the study area, flowing across Kulmstranda plain. The next river, the Lidelva, flows from Liddalen valley and along the southern edge of Hornsundneset plain to the sea. The two rivers are up to 0.5 m deep. Several small streams located between the two rivers flow from the slopes of the Struvefjella across the coastal plains to the sea. Their gradient is relatively large (ca. 100 m per 1.5–3.0 km on the plains). Both aforementioned plains are mostly dry, as they consist of rocky terraces with thin and fragmented marine gravel cover. Numerous shallow lakes (up to 1 m deep), permanent near the coast and seasonal closer to the mountains, have been created due to the damming of water by new and old raised coastal ridges, often found on the sandstone layers, which dip towards the interior of the land (Fig. 4). The exceptions are two small lakes in glacial troughs: Savitsjvatnet lake on the highest part of Kulmstranda plain and an unnamed lake with fish in the pass between the Sergeijevfjellet and Lidfjellet massifs.

Fig. 7. Outlet of a large underground karst channel on the Olsokflya plain coast with the Trollosen karst spring and one of the karst holes (1 m in diameter, enlarged in the upper photo and marked with a red arrow in the lower photo), which developed above the channel after 1986 as a result of intensifying karst processes. Photo: J. Dudek, 2008

The Slaklielva river flows through Slaklidalen valley located along the eastern boundary of the study area. The river splits up into two streams at the valley outlet and flows to the sea along the northern edge of Breinesflya plain, ca. 1 km south of Lidelva. Wiederfjellet mountain slopes are drained by other streams, which flow across the plain. Only Slaklielva stream and a stream flowing from Wiederdalen are supplied partly by small glaciers. All the streams are supplied by the active layer of permafrost because the plain is covered by thick fine-grained deposits. Numerous swamps and the slow outflow of water are caused by the small gradient of the plain (50 m per 3–4 km) and large areas of moss tundra (Fig. 5). The rivers are up to 1 m deep. There are several small lakes located more than 1.5 km from the sea in the southern part of the plain.

The wide braided channel of the Vinda glacial river cuts off the Breinesflya plain from the southeast. The Vinda river flows out of Bungebreen glacier's marginal zone. The largest river in western Sørkapp Land, the Bungeelva, flows out of Bungebreen glacier to the south and across its marginal zone and drains the Tørrflya coastal plain from the east.

There are several surprisingly small superficial streams, first of all pro-glacial streams, across the plain's narrow neck between Vitkovskibreen glacier's front moraine and the sea (0.5–1.0 km) behind the Bungeelva river. There are virtually no streams on the karst lowlands of Bjørnbeinflya and Olsokflya.

Water from Vitkovskibreen glacier and Hilmarfjellet massif flows via underground karst channels across the plain to the sea (Pulina 1977). There is a very large karst spring (several m^3/s), Trollosen, in the outlet of the biggest channel on the sea coast (Fig. 7).

There are two concentrations of thermo-mineral springs, up to 16.5°C, with H_2S and CO_2 (Werenskiöld 1920; Liestøl 1976; Pulina 1977; Krawczyk, Pulina 1991): in Bjørnbeinflya, where they supply a few small and very shallow lakes, and in Olsokflya. The small Luktvatnet lake (which means smelly), near the Olsokbreen's lateral moraine, may be supplied in a similar manner.

There are few melt-water (thaw) lakes and, closer to the glacier, pro-glacial streams in the Olsokbreen glacier's extensive marginal zone, in the southeastern corner of the study area.

Water circulation in the study area is described by Pociask-Karteczka (1990).

Michał Węgrzyn, Maja Lisowska, Maria Olech, Piotr Osyczka

VEGETATION

The flora of Svalbard is dominated by cryptogams, mainly lichenized fungi (lichens), 600 species, and bryophytes, 370 species (Elvebakk, Hertel 1996), and also cyanoprokaryota and algae (Matuła et al. 2007). Vascular plants are in the minority with about 170 species (Jónsdóttir 2005). Similar proportions can be observed in western Sørkapp Land, where lichen biota consist of 170 species (Olech 1990), while vascular plant flora includes 82 species (Dubiel 1990). This is typical for the High Arctic. Western Sørkapp Land is diverse in terms of habitat conditions such as terrain relief, geological subsoil, hydrological conditions, edaphic relationships, microclimate, etc. The vegetation in the region is dominated by lichens and bryophytes, and some flowering plants. Different vegetation types often create a complex mosaic.

The key factors determining the location of different types of vegetation in Arctic areas are terrain relief and hydrological conditions (Dubiel, Olech 1990; Elvebakk 1997).

Bedrock has an obvious influence on vegetation. The presence of Carboniferous sandstone correlates with the location of epilithic lichen communities and moss communities with *Racomitrium lanuginosum*, lichen tundra with *Flavocetraria nivalis* and *Cladonia rangiferina*. Triassic formations, in turn, overlap with vegetation dominated by vascular plants, e.g. *Saxifraga nivalis*, *Bistorta vivipara* (syn. *Polygonum viviparum*) and *Papaver dahlianum*. Other communities such as the *Gymnomitrion corraloides* community, the *Cetrariella delisei* community and the *Bistorta vivipara* community, are found on coastal ridges and small hills.

One of the most important environmental factors determining the structure and location of tundra vegetation is the availability of water in the substratum (Dubiel, Olech 1990; Elvebakk 1997). In extremely dry places, where the melting of snow is the principal source of water, and water is quickly removed, the following communities of epilithic lichens can be found: the *Orphniospora moriopsis* community, the *Umbilicaria arctica* community and the *Umbilicaria cylindrica* community. In addition, the lichen-dominated *Flavocetraria nivalis – Cladonia rangiferina* community and the *Cetrariella delisei* community are also found.

In depressions, often supplied with water by seasonal and summer-active streams, mire and wetland vegetation can be found. This includes grass and bryophyte-dominated communities such as the *Calliergon stramineum* community, the *Juncus biglumis* community, the *Dupontia pelligera – Arctophila fulva* community, the *Carex subspathacea* community and the *Saxifraga hyperborea – Ranunculus spitsbergensis* community.

Communities of ornithocoprophilous vegetation develop on mountain slopes or close to mountain slopes in the vicinity of large seabird colonies and in places visited by birds. These include the communities with *Phippsia algida*, the bryophyte-dominated *Tetraplodon mnioides* community as well as the community of crustose lichens with *Candelariella arctica*.

The tendency over the last few years has been to create a more general classification of communities and to present it on general maps. Two general vegetation maps containing information about the study area are available: (1) the Circumpolar Arctic Vegetation Map (CAVM Team 2003), and (2) the Vegetation Map of Svalbard (Elvebakk 2005). According to the Circumpolar Arctic Vegetation Map, northwest Sørkapp Land is located in bioclimatic Subzone B, the second-coldest of the five subzones in the Arctic (CAVM Team 2003). Vegetation in Subzone B is generally open or patchy (20–80% cover), with plants 5–10 cm tall, and a relatively high (30–60%) proportion of lichens and mosses, and this characteristic matches the vegetation in the study area quite well. Dry sites in northwest Sørkapp Land are dominated by the

Fig. 8. A herd of reindeer grazing on the lower slopes of Sergeijevfjellet. Damaged vegetation and a major change in slope micro-relief (appearance of so-called cattle terraces due to intensive trampling) are visible, apart from the solifluction stone stripes. Photo: M. Węgrzyn, 2008

zonal vegetation type of Subzone B, "prostrate dwarf-shrub, herb tundra" (P1 type). Another tundra-type characteristic for Subzone B, "sedge/grass and moss wetlands" (W1 type), is dominant at wet sites. A more detailed vegetation map of Svalbard by Elvebakk (2005) shows three major vegetation units in northwest Sørkapp Land: mesic, circumneutral tundra characterized by *Luzula nivalis*, mesic, acidic tundra characterized by *Luzula confusa* and *Deschampsia alpina* mires.

In polar terrestrial ecosystems, animals play a major role in the circulation of matter and in the flow of energy. Seabirds nesting sites in coastal areas are a particularly important part of Arctic ecosystems, as their presence determines the functioning of terrestrial biocenoses. Seabirds link terrestrial and marine ecosystems, as they feed in the sea and breed on land, and thus bring a considerable amount of organic matter, from sea to land, mainly in the form of guano. It is a very important source of nutrients for tundra vegetation, as an autochthonous source of biogenic substances, especially nitrogen and phosphorus, which are in short supply. Nutrient enrichment of areas adjacent to seabird colonies, which causes the stimulation of primary production and the development of luxuriant vegetation, is called the ornithocoprophilous effect (Eurola, Hakala 1977). Organic compounds found in guano are easily absorbed by plants.

Over the last few years, vegetation in the tundra of Sørkapp Land has been greatly influenced by herds of reindeer, which appeared in the 1990s. Reindeer forage on vascular plants, mainly graminoids and fruticose lichens, is causing significant changes in vegetation (Fig. 8).

ENVIRONMENTAL AND LANDSCAPE CHANGES

Wiesław Ziaja

CLIMATE WARMING

Changes in the climate in western Sørkapp Land mirror global climate fluctuations. This is especially true of the northern hemisphere and Spitsbergen. There is a lack of meteorological data to show continuous climate fluctuations for the entire Sørkapp Land peninsula. Indeed, basic meteorological observations (mainly temperature) were carried out in different places throughout the peninsula but for very short periods of time, mostly in the summer (Ziaja 1999). Continuous meteorological observations have been carried out at the Polish Polar Station, which belongs to the Institute of Geophysics of the Polish Academy of Sciences, on Isbjørnhamna bay (Marsz, Styszyńska 2007). The bay is located along the northern coast of Hornsund Fjord, only 10 km from the study area. Hence, the station's data may be considered representative of western Sørkapp Land.

The Little Ice Age ended with a very cold period in the 1890s in Spitsbergen, including Sørkapp Land. A relatively warm contemporary period "after the Little Ice Age" began in the early 20th century. Secondary cold and warm climate fluctuations occurred during the 20th and early 21st century. The most recent fluctuations, starting in the 1980s, appear to point to a significant warming trend (Brázdil 1988; Ziaja 2004). This is certainly true of the western island's coast, both north and south of Hornsund Fjord. The mean annual temperature increased by almost 2°C and the mean summer (July and August) temperature increased by 0.6°C since the 1980s when we began our research (Table 1).

A systematic temperature increase occurred during each month. The greatest warming occurred during a winter season from September to February. This statistically significant positive trend in the change in air temperature, also in June and

Table 1. Mean annual temperature changes at the Polish Polar Station at Hornsund, according to data from the Institute of Geophysics, Polish Academy of Sciences, published in part by Styszyńska (2007)

	1980–1989	**2000–2009**
Mean annual temperature	–5,2°C	–3,15°C
Mean summer, i.e. July and August, temperature	+3,9°C	+4,5°C

August (Marsz 2007), accelerated the thawing of snow and ice. This contributed to the lengthening of the vegetation season, the thawing of the permafrost active layer and the ablation of glaciers.

An increase in precipitation, both annual totals (statistically significant trend) and summer (July and August) totals has been detected since the 1980s. A clear increase in rain precipitation totals, the intensity of rainfall and the frequency of rainfall have been noted during this time period. This is also true of the summer season (Łupikasza 2007, Table 2). This increased rainfall intensified a number of geomorphic processes as well as snow and ice ablation. This, in turn, impacted the water cycle and vegetation.

Table 2. Mean annual changes in precipitation totals at the Polish Polar Station at Hornsund, according to data from the Institute of Geophysics, Polish Academy of Sciences, published in part by Styszyńska (2007)

	1980–1989	**2000–2009**
Mean annual sum of precipitation	390,5 mm	451,3 mm
Mean summer, i.e. July and August, sum of precipitation	78,1 mm	94,4 mm

Marsz and Styszyńska (2007) argue that the climate in southwestern Spitsbergen has changed rapidly and considerably during the past 30 years. The following are symptoms of this change: warming, an increase in oceanicity in terms of air temperature, an increase in precipitation totals, increasing share of rain in precipitation totals, a shortening of the snow season and an increase in cloudiness. This climate change prompted other changes as well.

Wiesław Ziaja, Justyna Dudek

GLACIAL RECESSION

Snow patches, permafrost and glaciers are the environmental components most sensitive to climate warming.

All the snow patches melt during the warmest and sunniest summer seasons across western Sørkapp Land lowlands and mountains, except for the patch, where the Lisbetelva river originates. In spite of climate warming, there are still summer seasons with numerous snow patches that last until the next winter season. Permafrost is more isolated from the summer heat (positive temperatures) by a layer of ground, which thaws down to a certain depth, i.e. the so-called active layer. In spite of the lack of a detailed survey, an increase in the active layer's thickness has been observed. Its maximum thickness in Quaternary deposits (apart from blocks and boulders) varied between 40 cm and 1.5 m, according to their granulation and vegetation, at sites below 100 m of altitude in 1986 (Ziaja 1988), and had doubled by in 2008. Rapid melting of thin (up to 2 m) ice lenticels occurs in rare pingo landforms in the lowest marine terrace (Fig. 9).

Glaciers react to warming with a few years (or longer) delay. Almost all Sørkapp Land glaciers have undergone a continuous recession since the beginning of the 20th century (Ziaja 1999; Błaszczyk et al. 2009). Only three small glaciers in the highest part of the Hornsundtind mountain group increased in thickness (and volume) from the 1930s to the 1980s (Ziaja 1999), but then became smaller as well. Two small glaciers, on the eastern slopes of Kovalevskifjellet and western slopes of Wiedefjellet, had already declined before the 1980s (Ziaja 1999). A few surges, i.e. rapid ice flows, resulted in temporary (up to several years) advances of a glacier terminus or in the thickening of a glacier tongue, could have occurred in western Sørkapp Land during the past century. Bungebreen glacier's surge after 1990 (Fig. 10) was described by Sund et al. (2009). Surges during a period of climate warming lead to the glaciers' recession because a downward shift of an ice mass results in the thinning of glaciers' firn fields, and thickened glacier tongues melt quickly, reducing glaciers' volume and surface area.

The following two processes are important for glacier recession in the study area: (1) a decrease in snow accumulation in firn fields due to the thawing of a larger snow mass in the summer, and (2) summer thawing of ice on the surface of glacier tongues, which results in a decrease in ice thickness in the tongues. Thus, the equilibrium-line altitude between the ablation zone and the accumulation zone of a glacier shifts

Fig. 9. Destruction of a small pingo (several m long, up to 1m thick) on the accumulation coast south of Palffyodden. Photo: J. Dudek, 2008

Fig. 10. Fragment of the Bungebreen glacier front seen (cross section) from the east, with characteristic ice stratification following the glacier's surge. Photo: J. Dudek, 2010

upward due to warming, which reduces the accumulation zone. Hence, the entire surface of the glaciers undergoes lowering each year, which results in a decrease in their volume and their overall retreat. The retreat occurs everywhere and not just at the front of the glacier. The thinning of glaciers around nunataks and in valleys with steep slopes causes the uncovering of rocky or rock-and-weathering slopes from underneath the ice. Large areas of slopes are visually reduced using a normal projection on a map. Frontal glacier retreat, which leads to the appearance of marginal zones, is most noticeable, as glaciers are least inclined in the front. However, freeing marginal zones of glaciers need not mean freeing them of ice. Marginal zones are predominantly formed out of dead (motionless) glacial ice covered with a thin (up to 2 m) and continuous layer of moraine or glacifluvial material. This dead ice successively thaws during the summer season. However, the thawing process is so slow that the volume of ice is still (after more than 100 years of glacial recession) larger than the volume of new moraine or glacifluvial deposits. Of course, landform and Quaternary deposits develop over time, as described next.

During the period of our study, since the 1980s, an acceleration of the glaciers' recession had occurred (Maps 1–4, Fig. 11). Why? First, because the glaciers in question are too large in relation to contemporary climate conditions and especially in relation to air temperature. Thus, the glaciers would shrink even after an end to climate warming. Second, since the 1980s, significant warming has additionally stimulated the set of the processes described above and positive feedbacks as well. New factors stimulating glacier recession have appeared, e.g. an increase in the share of rain in annual precipitation totals.

The glaciers' thickness generally decreased from 1990 to 2008, usually by 10–40 m, and in places, especially in the lower parts and edges of the glaciers, as much as 40–94 m. Olsokbreen, the largest glacier in terms of area and volume, has experienced the largest decrease in the thickness of ice (Fig. 12). A large decrease in the thickness of the upper parts of the glaciers is almost certain because the summer season of 2008 was exceptionally cloudy and not windy, and the winter snow mass did not melt as usual. This can be easily shown by comparing *TerraASTER* satellite images from 2007 and from 2008. Only in the lower part of Bungebreen glacier did its thickness increase by up to 20 m due to the aforementioned surge (Maps 3–4).

If one defines western Sørkapp Land as the region west of the glacial interior the recession of the glaciers leads to the expansion of western Sørkapp Land. The marginal zones of large glaciers – Gåsbreen (Figs. 11, 13), Bungebreen (Fig. 10) and Vitkovskibreen (Fig. 14) – have become considerably wider since the 1980s. The retreat or decline of the smaller glaciers between them have occurred as well. The recession of the glaciers changed how the environment and the landscape type of the two valleys along the eastern boundary of the study area function. The unnamed valley in the north, which falls towards Gåshamna bay, is at this point not dammed by Gåsbreen glacier (and was still dammed in 2000) and does not feature an ice-dammed lake (which did exist in 2000). Gåsbreen glacier has become thinner and its front has retreated 750 m since the 1980s (Fig. 11). Hence, the valley has become easily acces-

Fig. 11. Landscape changes in the Gåshamna area: the Goësvatnet glacial-dammed lake (which still existed in 2000) disappeared before 2005 because of the glacier's recession. Gåsbreen glacier is much shorter now and does not dam up the valley. View from the south. Upper photo: W. Ziaja. Lower photo: J. Niedźwiecki

Fig. 12. Inner Stormbukta bay with the new shoreline formed since the 1980s, seen from the slopes of Hilmarfjellet (altitude: 335 m). Photo: J. Dudek, 2008

sible: it is possible to follow its floor from the coast of Gåshamna to Slaklidalen valley and further to the south of the study area, since even the smaller glaciers in the area had disappeared or became smaller along with numerous snow patches. The front of Wiederbreen glacier (on the gentle eastern slopes of Widerfjellet) has retreated far enough to become invisible from the Breinesflya lowland. A significant retreat of Bungebreen glacier (almost 1 km) has occurred since the 1980s (Map 1, Fig. 10), in spite of the surge after 1990 (Sund et al. 2009). The low-situated tongue of this glacier became thicker temporarily due to the surge. The frontal retreat of Vitkovskibreen glacier (Fig. 14) was much slower: ca. 400 m since 1990 (Map 2). The glacier was concave and ended close (300–400 m) to the sea in the beginning of the 20th century, and then underwent a successive but moderate decrease in thickness.

The tidewater Olsokbreen glacier, which marks the end of the study area from the southeast, has undergone the largest (in the study area) recession after the Little Ice Age. A very large decrease in its thickness has resulted primarily in its frontal retreat, which caused Stormbukta (the largest bay along the study area coast) to appear. The Olsokbreen front has retreated 1–2 km since 1990 (Map 2, Fig. 12). The recession of the tidewater Körberbreen glacier, which marks the end of the study area from the northeast, is also noticeable but the degree of recession is much smaller (Map 1) due to its higher elevation and northern exposure.

Fig. 13. Landscape changes along a boundary between western Sørkapp Land and the peninsula's interior (with the Hornsundtind mountain group, 1431 m, in the background): Gåsbreen glacier's recession resulted in (1) great changes in terrain relief in valleys, (2) disappearance of the Goësvatnet glacial-dammed lake, and (3) formation of a new terraced river valley. View from the Savitsjtoppen mountain massif to the east. Upper photo: Pillewizer (1939). Lower photo: J. Dudek

Map 1 covers western Sørkapp Land without the vicinity of the Vitkovskibreen and Olsokbreen glaciers but with the western part of the peninsula's interior. In the area found on Map 1, the rate of the surface recession of glaciers was very similar in the 1980s and 1990s: 0.66–0.71 km^2, i.e. 0.6–0.7% of glaciated area per year. This rate has decreased somewhat since 2003. The share of the glaciated area covered by Map 1 decreased from almost 34% (106 km^2) in 1983–1984 to almost 29% (90.5 km^2) in 2007. A very small part (only 6%) of the area abandoned by glaciers was inundated by the waters of Hornsund Fjord: almost 1 km^2, i.e. 0.3% of the land area covered by Map 1 (Table 3).

Table 3. Changes in glaciation in the northwestern part of Sørkapp Land, covered by Map 1

<table>
<tr><th>Year</th><th>Land area
[km²]</th><th>Glaciated area
[km²
%]</th><th>Unglaciated area
[km²
%]</th><th>Land area decrease
[km²
%]</th><th>Rate of the glacial recession
[km²
%
of the glaciated area per year]</th><th>Areas abandoned by glaciers above the sea level
[km²
%]</th><th>Deglaciated areas transgressed by the sea
[km²
%]</th></tr>
<tr><td>1983–84</td><td>313.85</td><td>106.30
33.87</td><td>207.55
66.12</td><td rowspan="2">0.27
0.09</td><td rowspan="2">0.66
0.60</td><td rowspan="2">4.37
94.18</td><td rowspan="2">0.27
5.82</td></tr>
<tr><td>1990</td><td>313.59</td><td>101.66
32.41</td><td>211.92
67.59</td></tr>
<tr><td>1990</td><td>313.58</td><td>101.66
32.41</td><td>211.37
67.59</td><td rowspan="2">0.55
0.18</td><td rowspan="2">0.72
0.71</td><td rowspan="2">8.76
93.99</td><td rowspan="2">0.56
6.01</td></tr>
<tr><td>2003</td><td>313.04</td><td>92.34
29.50</td><td>220.71
70.50</td></tr>
<tr><td>2003</td><td>313.04</td><td>92.34
29.50</td><td>220.71
70.50</td><td rowspan="2">0.09
0.03</td><td rowspan="2">0.46
0.50</td><td rowspan="2">1.69
92.86</td><td rowspan="2">0.13
7.14</td></tr>
<tr><td>2007</td><td>312.95</td><td>90.52
28.93</td><td>222.42
71.07</td></tr>
<tr><td>1983–84</td><td>313.85</td><td>106.30
33.87</td><td>207.55
66.12</td><td rowspan="2">0.9
0.30</td><td rowspan="2">0.66
0.60</td><td rowspan="2">14.82
93.82</td><td rowspan="2">0.96
6.08</td></tr>
<tr><td>2007</td><td>312.95</td><td>90.52
28.93</td><td>222.42
71.07</td></tr>
</table>

Fig. 14. Vitkovskibreen glacier's front and marginal zone seen from the slopes of Hilmarfjellet (altitude: 330 m). The glacier was 2 km longer at the end of the Little Ice Age (in the 1890s). View from the southeast. Photo: J. Dudek, 2008

During the period 2003–2007 and the period 2007–2008, the very small areas abandoned by glaciers (0.219 km^2 in sum) became occupied by the Körberbreen and Bungebreen glaciers, which generally retreat rather quickly (Map 1). This episode does not affect the overall retreat of glaciers in western Sørkapp Land.

Map 2 covers the southeastern part of western Sørkapp Land in the vicinity of the Vitkovskibreen and Olsokbreen glaciers. In the area found on Map 2, the rate of the surface recession of glaciers between 1990 and 2007 was slightly slower than that in the northern part of the study area, i.e. 0.5% of glaciated area per year. However, the surface area in this case is actually much larger. This rate decreased to some extent after 2003, impacted by the same short-term climate fluctuation affecting the entire study area. The share of the glaciated area covered by Map 2 decreased from 61% (169 km^2) in 1990 to almost 57% (154.3 km^2) in 2007. A substantial part (almost 30%) of the ice-freed areas were transgressed by the waters of Stormbukta bay: 4.3 km^2, i.e. almost 2% of the land area covered by Map 2 (Table 4).

Changes in front extents of the biggest glaciers are shown in Table 5.

Table 4. Changes in glaciation in the southwestern part of Sørkapp Land, covered by Map 2

Year	Land area [km^2]	Glaciated area [km^2 %]	Unglaciated area [km^2 %]	Land area decrease [km^2 %]	Rate of the glacial recession [km^2 % of the glaciated area per year]	Areas abandoned by glaciers above the sea level [km^2 %]	Deglaciated areas transgressed by the sea [km^2 %]
1990	276.76	168.94 **61.04**	107.82 **38.96**	3.32 **1.20**	0.89 **0.52**	9.38 **74.03**	3.14 **25.07**
2004	273.43	156.42 **57.20**	117.00 **42.79**				
2004	273.43	156.42 **57.20**	117.00 **42.79**	1.30 **0.48**	0.70 **0.45**	0.94 **44.76**	1.16 **55.24**
2007	272.13	154.32 **56.71**	117.81 **43.29**				
1990	276.76	168.94 **61.04**	107.82 **38.96**	4.63 **1.68**	0.86 **0.51**	10.32 **70.59**	4.30 **29.41**
2007	272.13	154.32 **56.71**	117.81 **43.29**				

Table 5. Frontal retreat or advance of the big glaciers in western Sørkapp Land, according to Maps 1 and 2

	1983-84 – 1990	**1990–2003**	**2003–2007**	**2007–2008**
Körberbreen	110–180 m (retreat)	190–380 m (retreat)	10–70 m (advance)	0–60 m (advance)
Gåsbreen	70–150 m (retreat)	150–230 m (retreat)	300–450 m (retreat)	
Bungebreen	70–300 m (retreat)	220–600 m (retreat)	0–70 m (advance)	0–70 m (advance)
		1990–2004	**2004–2007**	**2007–2008**
Vitkovskibreen		0–370 m (retreat)	0–80 m (retreat)	
Olsokbreen		600–1200 m (retreat)	70–450 m (retreat)	10–370 m (retreat)

Wiesław Ziaja, Krzysztof Ostafin, Justyna Dudek

TRANSFORMATION OF LANDFORMS AND QUATERNARY DEPOSITS

The greatest changes in landforms and Quaternary deposits have occurred due to the glaciers' recession, as described above. New landforms have appeared. Accumulation landforms appeared in the front of glaciers and around glacier tongues in their marginal zones, i.e. on lowlands and valley floors abandoned by glaciers and in their fore-fields situated below the marginal zones (i.e. beyond the former reach of the glaciers, Figs 10–14). Erosion landforms, apart from pro-glacial river incisions (Fig. 15), prevail on the steep slopes of valleys and mountain massifs. Hence, the recession of the upper parts of glaciers results almost exclusively in erosion landforms, and the recession of the lower parts of glaciers leaves both erosion and accumulation landforms behind (with predominance of the latter).

The following new ice-free landforms have been observed: rocky and rock-and-weathering slopes, valley incisions (cut by glacial waters) of different types (gullies, gorges), and roche moutonnées. The decrease in glacier thickness uncovers nunatak slopes from top to bottom. The continuation of this process leads to the joining of nunataks into new mountain ridges or ranges. Once the ice is gone, these erosion landforms undergo mostly non-glacial denudation-erosion changes, especially if they are located on steep slopes and in significantly inclined upper parts of valleys. However, some of them, especially at the foot of the slope and in the lower parts of valleys, are being covered by new deposits, first of all moraine or glacifluvial deposits, but also talus, weathering and fluvial deposits.

Marginal zones began to appear in the front and around the lowest parts of glacier tongues in the beginning of the 20th century, when the glaciers were the widest (Maps 1, 2). These zones have been expanding due to frontal and lateral recession of glacier tongues ever since. Marginal zones develop from single ice-cored frontal moraine ridges into extensive complexes of different landforms, including intra-marginal sandurs with pro-glacial river beds, kettle holes, kames, eskers and low and narrow ridges of fluted moraines. Landforms of one type are transformed into landforms of another type, e.g. ice-cored moraine ridges into undulated moraine plains.

A particularly clear change was visible in the valley above the lowest part of Gåsbreen glacier due to its shortening in the period 2001–2004. The ice-dammed lake (Goësvatnet) was still there in 2000. However, in 2005, the lake did not exist, and

Fig. 15. Wide bed of the Vinda, which was a large pro-glacial river in the past (during the Little Ice Age) and declined due to Bungebreen glacier's recession. Photo: J. Dudek, 2008

its trough was in an initial stage of fluvial transformation (Ziaja, Ostafin 2007). By 2008, the terraced river bed built of alluvial deposits was formed, with waterfalls on rock outcrops (Figs 11, 13).

Landforms clearly stabilize in places where dead ice was never present or had thawed completely, which is expressed by a large decrease in the rate of landform evolution. This stabilization occurs in some of the smallest marginal zones and along edges of larger ones. However, dead ice is covered by superficial deposits in the majority of the marginal zones' area. The slow thawing of dead ice causes continuous geomorphic transformation and the lowering of marginal zones. The thawing of ice is stimulated by the influx of freshwater and seawater. This leads to the appearance of characteristic ephemeral forms of thaw holes and tongues of mud-flows (Fig. 16).

Of course, the cliffs of tidewater glaciers undergo the quickest of changes, i.e. retreat (Fig. 12). Stormbukta bay as a whole appeared due to the recession of Olsokbreen since 1900 (Wassiliew 1925). The new shore of this large bay is continuously transformed. The sea took 750 m of coastline and 11 ha of land along the northern shore of Stormbukta bay, and removed the coastline inland a distance of 5 to 250 m, mainly due to ablation of dead ice (Figs 12, 17) from 1990 to 2008. This process continues, which is evidenced by mud-flows from the lateral ice-cored moraine to the sea (Fig. 16). However, at a shorter distance to the glacier front, the land has widened by

Fig. 16. Mud-flow due to ablation of the dead-ice core of the Olsokbreen lateral moraine, on the northern Stormbukta coast. Photo: J. Niedźwiecki, 2008

Fig. 17. New northern coast of Stormbukta bay formed after the retreat of Olsokbreen glacier since the 1980s, due to ablation of dead-ice cores in the lateral moraine and sea action. Photo: J. Niedźwiecki, 2008

ca. 100–200 m (9 ha) due to marine accumulation of the former moraine and glacifluvial material from 1990 to 2008 (Map 2, Figs 12, 17, 18). The southern Stormbukta coast experiences the same processes. 2.3 ha was added to the land in a similar way on the eastern coast of the small bay at the front of Körberbreen glacier (Map 1).

The geomorphic transformation of areas devoid of glaciers in a warming climate is much slower (than described above) but noticeable. Changes in some parts of the shoreline are the most visible. Geomorphic sea action along the coast increased considerably (especially during storms) due to a significant shortening of the sea-ice season since the 1980s. Sea accumulation prevails in the bay south of the Palffyodden headland: the beach widens at a rate up to 1 m per year (Fig. 19). The reverse process, sea abrasion, acts on the Tørrflya plain built of Old-Holocene marine deposits: the beach coastline retreats up to 1 m per year, and the cliff behind the beach (cut from above during storms only) – up to 2 m per year (Fig. 20).

The reindeer population, regenerated in the 1990s, is a new geomorphic factor affecting large areas of western Sørkapp Land. There are simply too many reindeer, ca. 170 in 2008, because they are not threatened by any predators. They destroy vegetation and initiate or intensify erosion processes in many places by overgrazing, e.g. river erosion in the Hohenloheskardet valley and wind erosion on the Tørrflya (Fig. 21) and Bjørnbeinflya plains. Small dunes, which were overgrown and stabi-

Fig. 18. New northern coast of Stormbukta bay formed after the retreat of Olsokbreen glacier since the 1990s, due to sea action. Photo: J. Niedźwiecki, 2008

Fig. 19. Results of intensified geomorphic sea action in western Sørkapp Land during a few recent decades: the accumulation coast south of the Palffyodden headland. View to the north. Photo: J. Dudek, 2008

Fig. 20. Results of intensified geomorphic sea action in western Sørkapp Land during a few recent decades: the abrasion coast of the Tørrflya plain, ca. 9 m high, with an uncovered whalebone, in the right. Photo: J. Niedźwiecki, 2008

Fig. 21. Destruction of the small dunes, which is a result of damage to the vegetation cover due to overgrazing and trampling by reindeer on the Tørrflya coastal plain. Photo: J. Niedźwiecki, 2008

lized before (Gębica, Szczęsny 1988), have been made mobile by reindeer. Reindeer also impact microrelief on gentle slopes overgrown by tundra by trampling and thus forming so-called cattle terraces, e.g. at the foot of Sergeijevfjellet from the sea-side (Fig. 8).

Karst processes have intensified due to higher air temperatures and larger quantities of flowing water. Rapid development of karst holes was observed on the Olsokflya lowland. The holes were no deeper than 1 m during the summer of 1986. Some of the holes were 1.5–2.0 m deep in 2000, and at least 10–20 cm deeper in 2008. The enlargement of one of them resulted in the collapse of a cave (karst channel) roof just before Trollosen spring, which is a pro-glacial river (Fig. 7). The development of similar karst holes may also be observed above this karst channel.

Smaller changes in landforms include debris-flows on talus and talus-torrent fans as well as a halt in the development of nivation moraines.

Wiesław Ziaja, Krzysztof Ostafin

CHANGES IN WATER DRAINAGE AND NETWORKS

Climate warming has led to an increase in the quantity of liquid water in the Sørkapp Land natural environment due to more intensive ablation of snow patches and glaciers, and the deepening of the active layer (with ice) above permafrost.

This water flows to the sea in streams and rivers (up to 10 km long), under the ground in the active layer of the Quaternary deposits, and under the ground through rock fissures and karst channels. Surface and underground streams carry much more water today than in the 1980s. However, the soil and the surface are generally more dry in the lowland areas between the streams today due to the deepening of the active layer above the permafrost.

The river and lake network changed the most due to glacier recession.

Ice-dammed Goësvatnet lake disappeared due to the recession of Gåsbreen glacier. Its water table became lower and its surface decreased due to the thinning of the glacier tongue, which dammed the valley over time. However, the glacier did not become shorter and the lake persisted until 2000. After 2000, recession further shortened the glacier, removing the natural glacial dam, as detected in 2005 (Ziaja, Ostafin 2007). A large river flows from the upper part of the valley through the former lake's trough and is supplied below by water from Gåsbreen glacier (Figs 11, 12). A few smaller ice-dammed lakes disappeared as well.

However, other lakes have become larger due to glacier recession. This is especially true of a large lake dammed by a frontal moraine in the eastern part of Bungebreen glacier's marginal zone. A lot of smaller lakes appear, change their shape and disappear because of dead ice thawing in the larger glaciers' marginal zones.

Pro-glacial rivers change their course and flow both in the marginal zones of glaciers and their fore-fields, i.e. on the coastal lowlands. The greatest changes have occurred in the Bungebreen glacier fore-field. The previously large and rather short Vinda river has virtually disappeared due to the recession of Bungebreen glacier. Today, only small streams flow periodically in its large and deeply incised bed (Fig. 15). The Bungeelva river, which flows out of the aforementioned enlarged lake, became clearly larger, at least in summer.

Michał Węgrzyn, Maja Lisowska, Maria Olech, Piotr Osyczka

CHANGES IN VEGETATION

Plant communities in 1982–1985

The making of phytosociological maps of Arctic areas is difficult because of incomplete knowledge about tundra plant communities. One of the problems is mosaic vegetation made up of many different communities occupying small areas. Available vegetation maps usually show the distribution of physiognomic types of tundra (CAVM Team 2003) or the distribution of plant communities on a large scale (Elvebakk 2005).

A detailed vegetation map catalogues plant communities in a given area and shows their extent and geographic location. The phytosociological map of northwest Sørkapp Land (Dubiel, Olech 1991), which was the basis for the current comparative study, was made on a 1 : 25 000 scale. Mapping was preceded by making a series of relevés according to the classic Braun-Blanquet method and creating phytosociological tables enabling the identification of plant communities. Detailed phytosociological research resulted in the selection of 28 plant communities, which were clearly distinguished phytosociologically and were visible in the field. The aim of the research was not to define the precise syntaxonomic position and names of the communities in question. The results included a vegetation map, which would be the basis for further comparative studies on vegetation changes in the area, and the characterization of the plant communities identified during the research.

A group of plant communities associated with dry habitats was identified based on relief and water conditions. This group includes vegetation types, for which high water content in the substratum is not crucial for existence and which are relatively resistant to the drying properties of wind. This type of vegetation is dominated by bryophytes and lichens, which usually obtain water directly from the atmosphere. On the Kulmstranda and Hornsundneset plains, large areas of boulders and pebbles can be found, which are made of siliceous sandstone. Most were covered by communities of epilithic lichens. The *Orphniospora moriopsis* community occurred in places exposed to strong winds. A lichen-bryophyte community, represented by *Sphaerophorus globosus*, developed between stones. The floristically poor *Racomitrium lanuginosum* community was also found in the region in places covered with small rocks.

This community was found quite often in depressions between old storm ridges. The spatial formation of the dominant moss species *Racomitrium lanuginosum* followed the direction of local foehn winds (Fig. 22). In places which are protected from the wind, *Umbilicaria arctica* and *Umbilicaria cylindrica* communities achieved their optimum. In areas where the wind coming from Hornsund is halted by the slopes of Hohenlohefjellet, a moss and lichen-dominated vegetation complex occurred, composed of *Orphniospora moriopsis* and *Racomitrium lanuginosum* communities.

Much closer to the shoreline, on marine terraces built mainly of sea pebbles, a mosaic of *Cetrariella delisei* and *Gymnomitrion coralloides* communities (Fig. 23), which usually exist together, could be found. The *Cetrariella delisei* community in this formation was dominant in the southern part of the research area, while the *Gymnomitrion coralloides* community was found further north. Patches of this vegetation complex had a characteristic brown-greyish color. In the central part of Hornsundneset, full of rocky ledges near lakes, another type of vegetation mosaic was identified, composed of *Cetrariella delisei*, *Gymnomitrion coralloides* and *Racomitrium lanuginosum* communities.

Fot. 22. *Racomitrium lanuginosum* community on the upper marine terrace of the Hornsundneset coastal plain: a typical shape formed by foehn winds. Photo: M. Węgrzyn, 2008

Fig. 23. A mosaic composed of the *Cetrariella delisei* community and the *Gymnomitrion coralloides* community on the Hornsundneset coastal plain (with the Sergeijevfjellet and Lidfjellet mountain massifs in the background). Fissures of frost polygons are visible. Photo: J. Dudek, 2008

Another mosaic dominated by the *Cetrariella delisei* community was one composed of *Cetrariella delisei*, *Gymnomitrion coralloides* and *Orphniospora moriopsis* communities. This type of formation, occupying mainly rocky ledges near the sea in an area of strong winds, was relatively rare.

The most spectacular vegetation type in the study area was most likely the *Flavocetraria nivalis – Cladonia rangiferina* community. The principal component of this community included numerous species of fruticose epigeic and epibryophytic lichens, along with bryophytes (Fig. 24). The *Flavocetraria nivalis – Cladonia rangiferina* community could be found on initial soils in areas protected from the wind by mountain slopes (Fig. 25). The best developed patches of this community were noted at the foot of Hohenlohefjellet, while a less developed form was found in the central part of the coastal plain near the sea and lakes. The *Flavocetraria nivalis – Cladonia rangiferina* community and the *Cetrariella delisei* community and the *Gymnomitrion coralloides* community form a complex found on rocky ledges between lakes.

All the vegetation types described above occupied the northern part of the research area. Farther south, a larger share of vascular plants was observed. Vascular plant-dominated communities were the most floristically rich in the study area. Their distribution was associated with areas of moderate soil humidity.

Fig. 24. General view of tundra vegetation in the central part of western Sørkapp Land (the Hornsundneset coastal plain and the Sergeijevfjellet mountain massif) in 1982. In the foreground: the *Carex subspathacea* community; in the background – the *Flavocetraria nivalis* – *Cladonia rangiferina* community with visible patches of yellow fruticose lichens. Photo: M. Olech

Fig. 25. *Flavocetraria nivalis* – *Cladonia rangiferina* community at the foot of Hohenlohefjellet. Photo: M. Olech, 1982

The *Saxifraga aizoides* community occupied large parts of the Breinesflya plain. It is a place which is annually flooded by water from melting snow. Therefore, a large amount of water is available only at the beginning of the vegetation season. Stony soil in this area includes a large amount of humus and a lot of calcium carbonate. The *Saxifraga aizoides* community was one of the most floristically rich communities in the study area.

Like the previous plant community, the *Bistorta vivipara* community also was floristically rich. It developed in the lower parts of the slopes of Sergeijevfjellet and Lidfjellet, as well as on elevated marine terraces, which are very favorable for plant development. The *Bistorta vivipara* community also formed a species-poor complex of vegetation with the *Cetrariella delisei* and the *Gymnomitrion coralloides* communities. This complex was found at numerous locations on marine terraces of Breinesflya and on the slopes of Kvartsittrabben.

The *Saxifraga nivalis* community was often found on moist, western mountain slopes. The substratum in these areas is fine-grained with signs of solifluction. This community was rich in species of both vascular plants and cryptogams – bryophytes and lichens.

The *Papaver dahlianum* community developed on scree slopes with western exposure, and on the gravelly sides of seasonal streams (Fig. 26). This community was floristically poor, mainly due to unstable ground.

The *Juncus biglumis* community was dominant in the southern part of the research area, i.e. on the Breinesflya plain, on the slopes of Lidfjellet and Wiederfjel-

Fig. 26. *Papaver dahlianum* with very expansive *Salix polaris* in the *Papaver dahlianum* community. Photo: M. Węgrzyn, 2008

let. Large patches of this community were found in places with long lasting snow cover and frequent solifluction episodes. The ground in these areas is moist during the entire vegetation season.

Rather infrequent in the study area, the *Saxifraga oppositifolia* community occupied seashore areas with poor gravelly-sandy soil.

Another group of plant communities was associated with wet areas such as land depressions, stream beds and lake shores. They were generally found in the central and southern parts of the research area.

A mosaic composed of the *Dupontia pelligera* community and the *Arctophila fulva* community appeared along streams and on moist lake shores. The first community preferred a very wet habitat, while the second grew in less moist places. This complex included numerous graminoids and resembled meadow formations.

The *Carex subspathacea* community developed in the wettest areas, often flooded during snowmelt season, and also by permanent and seasonal streams (Fig. 24). Under a layer of vascular plants and bryophytes, a 15 cm thick layer of peat was formed on gravelly or gravelly-sandy substratum. The largest areas of the *Carex subspathacea* community were found usually far from the sea, in land depressions at the foot of Lidfjellet and Sergeijevfjellet.

A typical peat-producing vegetation type in Sørkapp Land was the *Saxifraga hyperborea – Ranunculus spitsbergensis* community. This very floristically poor community formed in boggy areas on silt or sand with the addition of gravel. The layer of peat underneath the plants was 20–80 cm thick. Vascular plants typical of this community occurred infrequently within thick patches of mosses, mainly *Calliergon stramineum*.

Gravelly areas close to streams running across coastal plains were occupied by a very species-poor *Calliergon stramineum* community. The dominant species was a moss *Calliergon stramineum*. During the summer, areas occupied by these communities are often flooded by the changing water level of the streams. In the tundra landscape, one can easily spot those deep green patches among vegetation in dry habitats, which is grey or in various shades of brown (Fig. 24). In larger land depressions, a similar vegetation type can be found, i.e. the *Calliergon sarmentosum* community.

Ornithocoprophilous vegetation, associated with seabird colonies, is completely different from the previous types. Soil constantly fertilized by guano rich in nutrients, i.e. compounds of phosphorus and nitrogen, is the key habitat-creating factor. In the study area, large nesting colonies of the Little Auk *Alle alle* found on the slopes of Hohenlohefjellet and Sergeijevfjellet fertilize the nearby area.

Ornithocoprophilous vegetation, which developed directly in bird colony areas included the *Candelariella arctica* community, composed of epilithic lichens with predominant yellow-colored *Candelariella arctica*. It developed on the surfaces of rocks and boulders of the Hohenlohefjellet and Sergeijevfjellet slopes.

The *Tetraplodon mnioides* community, on the other hand, developed below bird colonies. It benefits from compounds of phosphorus and nitrogen washed out from bird colonies and carried by water into streams. This community, composed entirely of the ornithocoprophilous bryophytes *Tetraplodon mnioides* and *Aplodon wormskjoldii*, was found at the base of the mountains, where the soil was soppy.

Apart from all the plant formations mentioned above, one could also find small areas almost entirely devoid of vegetation, where only initial communities of epilithic lichens have developed. These places were located in high mountain areas covered by snow most of the year.

Plant communities in 2008 – changes over the last 25 years

The analysis of vegetation in Sørkapp Land done in 1982 and 1985 (Dubiel, Olech 1990, 1991) was repeated in 2008. Historical phytosociological data and maps were compared with new results.

Phytosociological methodology is very effective for the monitoring of the state of vegetation and for observing changes in the composition and distribution of different vegetation types over time. Relevés made in the same area as during the previous study (Dubiel, Olech 1990, 1991) and a repeat of vegetation mapping allowed for an analysis of changes in vegetation in the area.

Vegetation dynamics in the research area and trends, in the context of factors operating during the last 25 years and currently, were analyzed.

On the basis of repeated vegetation mapping, significant changes in composition and extent of several plant communities were documented. The principal trend is a decrease in species diversity, leading to a more uniform vegetation. This process is most visible in dry habitats, while in wet areas, changes are less noticeable.

One of the most spectacular changes was the complete degradation of the *Flavocetraria nivalis – Cladonia rangiferina* community at the foot of Hohenlohefjellet and Sergeijevfjellet. This community, previously dominated by fruticose lichens from the genus *Cladonia* and *Cetraria*, has transformed completely. Currently, *Flavocetraria nivalis*, *Cladonia rangiferina*, and other species of *Cladonia* have practically disappeared. The moss *Racomitrium lanuginosum* and the lichen *Sphaerophorus globosus* became dominant instead. A larger number of vascular plants, especially *Luzula arcuata*, was also recorded. It seems that the *Flavocetraria nivalis – Cladonia rangiferina* community transformed into a moss-dominated one, with a predominance of *Racomitrium lanuginosum* (Fig. 27).

The main component of the *Flavocetraria nivalis – Cladonia rangiferina* community, fruticose lichens, was entirely consumed by reindeer. The remaining lichens gradually declined due to the fertilization of the substratum. Currently, these lichen species can only be found on very steep slopes of Hohenlohefjellet and in rock crevices on coastal plains, in places inaccessible to reindeer.

The patches of the *Flavocetraria nivalis – Cladonia rangiferina* community, which used to form a mosaic together with the *Cetrariella delisei* and the *Gym-*

Fig. 27. Degradated form of the *Flavocetraria nivalis – Cladonia rangiferina* community as a result of reindeer grazing, trampling and fertilizing over the last few decades, in the uppermost part of Hornsundneset at the foot of Hohenlohefjellet. Photo: M. Lisowska, 2008

nomitrion coralloides communities, were also destroyed. Currently, only *Cetrariella delisei* and *Gymnomitrion coralloides* are present in these areas.

A similar trend of unification is visible in places, where a mosaic of the *Racomitrium lanuginosum* and the *Orphniospora moriopsis* communities used to exist. Presently, boundaries between these two communities have become unclear as a result of the expansion of *Racomitrium lanuginosum*. This species used to occupy depressions between storm ridges, where fine-grained rock material would accumulate. Now its extent has moved onto the top parts of the ridges.

Within vegetation complexes with a predominance of the *Cetrariella delisei* community, no clear boundaries between the complex components currently exist. *Cetrariella delisei* has expanded greatly along the shore, just like *Racomitrium lanuginosum* has done inland.

The *Cetrariella delisei* community creates a well-defined complex with the *Gymnomitrion coralloides* community. Within this mosaic, the expansion of the *Cetrariella delisei* community is also visible.

In the central part of Hornsundneset, an earlier map showed a complex of the *Cetrariella delisei*, the *Gymnomitrion coralloides*, and *Racomitrium lanuginosum* communities, however currently the *Cetrariella delisei* has expanded. Boundaries between the components have vanished, which reflects the general trend in the area, and the same situation is visible in the complex of the *Cetrariella delisei* and the *Orphniospora moriopsis* communities. In the latter complex, stony ledges occupied previously by *Orphniospora moriopsis* and other epilithic lichens, are comparatively small patches in an area becoming more and more dominated by fruticose lichens.

Changes in the structure of plant communities in wet areas are not as substantial as in communities associated with dry habitats.

The structure of the *Saxifraga aizoides* community, which covers large areas of the Breinesflya plain, has not changed since the 1980s.

In the *Bistorta vivipara* community, which is also floristically rich, an increase of abundance of *Salix polaris* was recorded. It is clear that *Salix polaris* is an expansive species in this community.

On the shore, in the *Cetrariella delisei* and the *Gymnomitrion coralloides* communities, expansion of *Salix polaris* was not observed. Consequently, the expansion of this species may be directed from the shore inland, towards mountain slopes.

The dynamics of the *Saxifraga nivalis* community is stable, and no changes in the structure of the community were observed. However, within the *Papaver dahlianum* community, which occupies the western slopes of the mountains, an expansion of *Salix polaris* is evident (Fig. 26). In the 1980s, *Salix polaris* used to be found in this community only occasionally. However, now it grows in dense patches.

The occurrence of *Luzula arcuata* was noted in this community for the first time.

Within the *Juncus biglumis* community, which dominates in the southern parts of the research area, on the Breinesflya plain and at the foot of Lidfjellet and Wiederfjellet, the structure remains unchanged. Both the boundaries of the formation and the species structure seem to be stable.

The *Saxifraga oppositifolia* community was very diverse in the 1980s. Presently, major changes can be observed within this pioneer community. The previously dominant *Saxifraga oppositifolia* is in decline, while other expansive species, which used to be scarce, appear in larger quantities. *Salix polaris* is found in large quantities, as is *Cetrariella delisei*, which migrates from neighboring communities. *Gymnomitrion coralloides* can be observed to a smaller extent as well. On the updated vegetation map, the *Saxifraga oppositifolia* community includes a mosaic composed of the *Cetrariella delisei* and the *Gymnomitrion coralloides* communities.

In permanently wet areas, the extent of the following communities did not change: the *Dupontia pelligera* and the *Arctophila fulva* communities, the *Carex subspathacea* community, the *Saxifraga hyperborea – Ranunculus spitsbergensis* community (Fig. 28), the *Calliergon stramineum* community and other moss bog variants. Species composition also does not fundamentally deviate from the one noted in the 1980s. Only the general condition of plants and the state of preservation have visibly changed due to reindeer grazing.

Within ornithocoprophilous vegetation, species composition remained the same and changes in their extent were not observed. This is associated with the specific character of these communities, one composed of epilithic lichens, with leading spe-

Fig. 28. *Saxifraga hyperborea – Ranunculus spitsbergensis* community on the Hornsundneset plain just below Hehenloheskardet. Photo: M. Węgrzyn, 2008

cies *Candelariella arctica*, and the other typically bryophytic, created mainly by *Tetraplodon mnioides*. The size of the colonies of *Alle alle* seems to be unchanged, therefore ornithocoprophilous vegetation constantly receives enough nutrients to prevail.

Communities of epilithic lichens, with the *Umbilicaria arctica* and the *Umbilicaria cylindrica* communities, are rather stable and have not transformed much over the last 25 years. Patches of these communities are found in the northern part of the study area, where extensive rock surfaces, boulders and storm ridges made of sea pebbles occur frequently.

Causes of changes

Vegetation maps are a valuable source of information for estimating changes in vegetation in a given area. They are also very useful for assessing the speed and direction of changes, which is particularly important because vegetation is dynamic, influenced by numerous factors such as climate changes, animal and human impact.

These factors and the changes caused by them have been thoroughly analyzed throughout the Arctic. The research results provide insight into the speed of the changes and their direction, and allow to make predictions concerning future trends.

Svalbard ecosystems are described in the literature as adapted to extreme fluctuations in climate conditions on different temporal scales and slowly reacting to climate changes (Jónsdóttir 2005). Research confirms this, revealing that the impact of climate change on vegetation can not be detected in whole Svalbard. Vegetation mapping in Adolfbukta (Billefjorden, middle Spitsbergen) repeated in 2008 after 70 years did not show any changes in vegetation (Prach et al. 2010). The authors suggest that the reason for such a state of affairs, apart from the Arctic ecosystems' slow reaction to climate changes, could be the pressure of herbivores, masking the possible development of vegetation in that area.

In the case of Sørkapp Land, research results show a different situation. Major changes in vegetation have taken place since the 1980s, especially in dry or slightly moist areas. On the contrary, minute changes were recorded in very wet areas. The results of experimental research in Central Spitsbergen (Speed et al. 2010) were in line with observations from Sørkapp Land. A herbivore simulation (goose pressure) showed that different vegetation types have different levels of resistance to herbivore disturbance. Wetlands and mires proved to be the most resistant.

The main cause of vegetation changes in Sørkapp Land is the rapidly growing population of reindeer in the area. In addition to environmental changes favoring reindeer expansion, the legal protection of these animals in the area was also of great importance thanks to the creation of South Spitsbergen National Park in 1973. A rapid expansion of the reindeer population was observed, from 1 or 2 animals in the 1980s to ca. 100 in the year 2000 (Ziaja 2002), and even later to about 170 animals in 2008. The growing number of reindeer has a strong effect on the structure and distribution of plant communities. The impact of herbivores has included grazing on lichens and over-ground parts of vascular plants, as well as the trampling of bogs (Fig. 29) and the fertilization of the ground.

Svalbard reindeer, unlike other reindeer subspecies, hardly migrate (Tyler, Øritsland 1989). It can thus be assumed, that in the future the population of reindeer in western Sørkapp Land will not decrease, provided there is enough forage.

For comparison, the reintroduction of reindeer in the area of Ny Ålesund, Northwest Spitsbergen, caused a major transformation of the tundra over a short period of time, and the effects are similar to those observed in Sørkapp Land. In 1978 twelve reindeer were brought to Brøggerhalvøya, near Ny Ålesund. Until 1986–1987, the number of animals increased to 110 (Øritsland 1987). In 1991 there were already 300 reindeer (Wegener et al. 1992, after: Elvebakk 1997). Only one year after the introduction of reindeer, the destruction of the *Flavocetraria nivalis* population in Brøggerhalvøya was observed, and one year after that, *Cladonia arbuscula ssp. mitis* disappeared as well (Elvebakk 1997). After some time, it was observed that a part of the *Cetrariella delisei* population, previously not grazed on by reindeer, was also destroyed (Elvebakk 1997). Negative changes in the population of epigeic fruticose li-

1983

2008

chens can be also observed in other regions of the Arctic, for instance in Alaska (Joly et al. 2009), where a decrease in cover and biomass of ground lichens was recorded in recent years. This tendency is associated with several factors such as reindeer and caribou activity as well as competition with vascular plants spreading as a result of climate change. An experiment simulating the influence of global warming on tundra resulted in a decrease in lichen cover (Hudson, Henry 2010).

◀

Fig. 29. Wet moss tundra in the outlet of Hohenloheskardet valley at a time without any reindeer in the upper photograph, and the same tundra with a visible effect of reindeer grazing and trampling since the 1990s in the bottom photo. Top photo: W. Ziaja. Bottom photo: M. Węgrzyn

CONCLUSIONS AND PROGNOSIS FOR ENVIRONMENTAL CHANGE

Wiesław Ziaja

LANDSCAPE DEVELOPMENT

A clear climate change, reflected in a significant increase in air temperature and atmospheric precipitation, is the reason for an increase rate of landscape development. Therefore, each environmental-landscape component, apart from solid bedrock, has changed.

The first and direct result of climate warming has been glacial recession, which stimulated an entire process of landscape (and seascape) changes along the eastern boundary between western Sørkapp Land (devoid of glaciers during the Holocene) and the glaciated peninsula's interior. These profound landscape changes, described above, have mainly been the result of the after-effects of the glaciers' retreat. A completely new landscape has appeared.

The fore-fields of glaciers, i.e. areas outside the glaciers' maximum extent, have been indirectly influenced by the glaciers' retreat. Landforms, Quaternary deposits and water networks have been significantly affected, especially on the coastal plains in the north and south of the study area: Gåshamnøyra and Tørrflya.

Some sequences of non-glacial and non-postglacial coastline have been affected by an increase in the geomorphic activity of the sea due to a shorter sea-ice season, which can be observed via a clear retreat (under abrasion) or advance (under accumulation) of the coastline. The formation of new coastlines at the front of the tidewater glaciers (Körberbreen and Olsokbreen) is a more complicated process.

Current environmental and landscape changes in large parts of the coastal lowlands and mountains – being ice-free during the Holocene – are much more successive and not as explicit as changes previously mentioned. However, a longer summer season and less snow patch persistence, an increased quantity of water in the ground surface and the active layer of permafrost affected geomorphic and hydrologic processes everywhere. After 25 years, this effect is visible in a slower rise of nivation moraines, more frequent debris-flows and more intensive fluvial processes.

During the next few decades, the described trend of environmental-landscape transformation will continue unless the climate cools down. In the case of a progressive warming, the extensive tongues of the Gåsbreen, Bungebreen and Vitkovskibreen glaciers will first retreat and then disappear. The principal reason for this is that the main part of each glacier is located at an altitude of 300 m above sea level or less. The same is true of low-lying smaller glaciers or their parts. The main result of that would be an expansion of non-glacial landscape, vegetation and animal life to the east, into the currently glaciated peninsula's interior.

Michał Węgrzyn, Maja Lisowska, Maria Olech, Piotr Osyczka

VEGETATION CHANGES

On the basis of tundra vegetation mapping repeated after 25 years, major changes in the structure of several plant communities and their extent were recorded. First of all, a decrease in species diversity is visible, leading to more homogenous vegetation. This process is more noticeable in dry areas, whereas in wet areas, the changes are smaller. One of the most spectacular changes was a complete degradation of the fruticose ground lichen population, mainly in the *Flavocetraria nivalis – Cladonia rangiferina* community.

It seems necessary to try to predict the direction of future vegetation changes in Sørkapp Land. It is probable that the reindeer population in western Sørkapp Land will not decrease over the next few years. Climate change scenarios indicate a high probability of further warming of the climate.

The results of research from other parts of the Arctic can provide necessary information for estimating future changes. It is assumed that in the areas grazed by reindeer, graminoids usually increase (Elvebakk 1997). This is also true in lichen-dominated plant communities (Klein 1968; Post, Klein 1999) and they are also predicted to increase under global warming scenarios (Walker et al. 2006). On the other hand, macrolichens, although important for the functioning and biodiversity of cold northern ecosystems, are predicted to be negatively affected by climate change (Cornelissen et al. 2001). Therefore, the increase of graminoids in the coming years is likely to take place in the research area.

The degree of preservation of fruticose lichens may be treated as a good indicator of changes taking place in the presence of a large population of reindeer. It can be also, indirectly, used to determine the level of negative effects of climate change in the Arctic (Cornelissen et al. 2001; Walker et al. 2006; Joly et al. 2009).

References

Błaszczyk M., Jania J.A., Hagen J.O. (2009): Tidewater glaciers of Svalbard: Recent changes and estimates of calving fluxes. *Polish Polar Research*, 30(2): 85–142.

Brázdil R. (1988): Variation of air temperature and atmospheric precipitation in the region of Svalbard [in:] Results of Investig. of the Geogr. Research Expedition Spitsbergen 1985. *Folia Facultatis Scientiarum Naturalium Universitatis Purkynianae Brunensis* (Brno), *Geographia*, 24: 285–323.

Braun-Blanquet J. (1964): Pflanzensoziologie: Grundzüge der Vegetationskunde. 3. Auflage. Springer Verlag, Wien–New York, 865 pp.

CAVM Team (2003): Circumpolar Arctic Vegetation Map. Scale 1 : 7,500,000. Conservation of Arctic Flora and Fauna (CAFF) Map No. 1. U.S. Fish and Wildlife Service, Anchorage, Alaska.

Cornelissen J.H.C., Callaghan T.V., Alatalo J.M., Michelsen A., Graglia E., Hartley A.E., Hik D.S., Hobbie S.E., Press M.C., Robinson C.H., Henry G.H.R., Shaver G.R., Phoenix G.K., Gwynn Jones D., Jonasson S., Chapin III F.S., Molau U., Neill C., Lee J.A., Melillo J.M., Sveinjornsson B., Aerts R. (2001): Global change and arctic ecosystems: is the lichen decline a function of increases in vascular plant biomass? *Journal of Ecology*, 89(6): 984–994.

Czeppe Z., Ziaja W. (1985): Structure of the geographical environment of the north-western Sörkappland (Spitsbergen). *Zeszyty Naukowe Uniwersytetu Jagiellońskiego, Prace Geograficzne*, 63: 19–32.

Dallmann W.K., ed. (1999): *Lithostratigraphic Lexicon of Svalbard. Upper Palaeozoic to Quaternary bedrock. Review and recommendations for nomenclature use.* Norsk Polarinstitutt, Tromsø, 318 pp.

Dallmann W.K., Birkenmajer K., Hjelle A., Mørk A., Ohta Y., Salvigsen O., Winsnes T.S. (1993): Geological Map, Svalbard, 1 : 100,000, C13G, Sørkapp (text). *Norsk Polarinstitutt Temakart*, 17, 73 pp.

De Geer G. (1923): Pl. F. Environs de la Station Russe D'Hivernage. 1 : 50000 (map). [in:] *Mesure D'un Arc de Méridien au Spitzberg, Entr. en 1899–1902, Description Topographique de la Région Explorée. Géologie*, Stockholm, Aktiebolaget Centraltryckeriet.

Dubiel E. (1990): Vascular plants of the NW Sörkapp Land (Spitsbergen). Distribution and habitats. *Zeszyty Naukowe Uniwersytetu Jagiellońskiego, Prace Botaniczne*, 21: 7–33.

Dubiel E., Olech M. (1990): Plant Communities of NW Sörkapp Land (Spitsbergen). *Zeszyty Naukowe Uniwersytetu Jagiellońskiego, Prace Botaniczne*, 21: 35–74.

Dubiel E., Olech M. (1991): Phytosociological map of NW Sörkapp Land (Spitsbergen). *Zeszyty Naukowe Uniwersytetu Jagiellońskiego, Prace Botaniczne*, 22: 47–54.

Elvebakk A. (1997): Tundra diversity and ecological characteristics of Svalbard [in:] F.E. Wielgolaski (ed.): *Polar and alpine tundra. Ecosystems of the World, 3.* Elsevier, Amsterdam: 347–359.

Elvebakk A. (2005): A vegetation map of Svalbard on the scale 1 : 3,5 mill. *Phytocoenologia*, 35(4): 951–967.

Elvebakk A., Hertel H. (1996): Part 6. Lichens [in:] A. Elvebakk A., Prestrud P., eds.: *A catalogue of Svalbard plants, fungi, algae and cyanobacteria. Norsk Polarinstitutt Skrifter*, 198: 271–359.

Eurola S., Hakala A.V.K. (1977): The bird cliff vegetation of Svalbard. *Aquilo, Series Botanica*, 15: 1–18.

Gullestad N., Klemetsen A. (1997): Size, age and spawning frequency of landlocked arctic charr *Salvelinus alpinus* (L.) in Svartvatnet, Svalbard. *Polar Research*, 16(2): 85–92.

Gębica P., Szczęsny R. (1988): Symptoms of aeolian accumulation in western Sørkapp Land, Spitsbergen. *Polish Polar Research*, 9(4): 447–460.

Hisdal V. (1985): Geography of Svalbard. *Polarhåndbook*, 2. Norsk Polarinstitutt, Oslo, 81 pp.

Hudson J.M.G., Henry G.H.R. (2010): High Arctic plant community resist 15 years of experimental warming. *Journal of Ecology*, 98: 1035–1041.

Hutchinson M.F. (1989): A new procedure for gridding elevation and stream line data with automatic removal of spurious pits. *Journal of Hydrology*, 106: 211–232.

Joly K., Jandt R.R., Klein D.R. (2009): Decrease of lichens in Arctic ecosystems: the role of wildfire, caribou, reindeer, competition and climate in north-western Alaska. *Polar Research*, 28: 433–442.

Jónsdóttir I.S. (2005): Terrestrial Ecosystems on Svalbard: Heterogeneity, Complexity and Fragility from an Arctic Island Perspective. *Biology & Environment: Proceedings of the Royal Irish Academy*, 105(3): 155–165.

Kalicki T. (1985): The foehnic effects of the NE winds in Palffyodden region (Sörkappland). *Zeszyty Naukowe Uniwersytetu Jagiellońskiego, Prace Geograficzne*, 63: 99–106.

Klein D.R. (1968): The introduction, increase, and crash of reindeer on St. Matthew Island. *Journal of Wildlife Management*, 32: 350–367.

Korona J., Berthier E., Bernard M., Remy F., Thouvenot E. (2009): SPIRIT. SPOT 5 stereoscopic survey of Polar Ice: Reference Images and Topographies during the fourth International Polar Year (2007–2009). *ISPRS Journal of Photogrammetry and Remote Sensing*, 64: 204–212.

Krawczyk W.E., Pulina M. (1991): Thermokarstic and glaciokarstic hydrology in Spitsbergen glaciers. *Proceedings, 1st International Symposium of Glacier Caves and Karst in Polar Regions*, Madrid: 185–198.

Kuczek M., Ziaja W. (1990): The structure of the geographical environment of the area between two glaciers: Vitkovskibreen and Olsokbreen. *Zeszyty Naukowe Uniwersytetu Jagiellońskiego, Prace Geograficzne*, 81: 31–56.

Leszkiewicz J. (1982): Źródła krasowe pod Rasstupet na południowym Spitsbergenie w świetle badań z 1978 roku. *Wyprawy Polarne Uniwersytetu Śląskiego 1977–1980*, tom I: 77–86.

Liestøl O. (1976): Pingos, springs, and permafrost in Spitsbergen. *Norsk Polarinstitutt Årbok*, 1975: 7–29.

Lindner L., Marks L. (1993): Middle and Late Quaternary evolution of the Hornsund Region, South Spitsbergen. *Polish Polar Research*, 14(3): 85–142.

Łupikasza E. (2007): 11. Opady atmosferyczne [in:] A.A. Marsz, A. Styszyńska (eds.): *Klimat rejonu Polskiej Stacji Polarnej w Hornsundzie – stan, zmiany i ich przyczyny.* Akademia Morska w Gdyni, Uniwersytet Śląski, Instytut Geofizyki PAN, Gdynia: 185–196.

Marsz A.A. (2007): 9. Temperatura powietrza: 9.1. Roczna temperatura powietrza. 9.2. Miesięczna temperatura powietrza. 9.3. Przebieg roczny temperatury dobowej. 9.4. Sezony termiczne [in:] A.A. Marsz, A. Styszyńska (eds.): *Klimat rejonu Polskiej Stacji Polarnej w Hornsundzie – stan, zmiany i ich przyczyny.* Akademia Morska w Gdyni, Uniwersytet Śląski, Instytut Geofizyki PAN, Gdynia: 131–146.

Marsz A.A., Styszyńska A., eds. (2007): *Klimat rejonu Polskiej Stacji Polarnej w Hornsundzie – stan, zmiany i ich przyczyny.* Akademia Morska w Gdyni, Uniwersytet Śląski, Instytut Geofizyki PAN, Gdynia, 376 pp.

Matuła J., Pietryka M., Richter D., Wojtuń B. (2007). Cyanoprokaryota and algae of Arctic terrestrial ecosystems in the Hornsund area, Spitsbergen. *Polish Polar Research*, 28(4): 277–282.

Nuth C., Kääb A. (2011): Co-registration and bias corrections of satellite elevation data sets for quantifying glacier thickness change. *The Cryosphere*, 5: 271–290.

Olech M. (1990): Lichens of the NW Sörkapp Land (Spitsbergen). *Zeszyty Naukowe Uniwersytetu Jagiellońskiego, Prace Botaniczne*, 21: 197–210.

Øritsland N.A. (1987): The reindeer in the Ny-Ålesund area. *Norsk Polarinstitutt Årbok*, 1986–1987: 38–40.

Pillewizer W. (1939): Die Kartographischen und Gletscherkundlichen Ergebnisse der Deutschen Spitsbergenexpedition 1938. *Petermanns Geographische Mitteilungen*, 238, 46 pp. + 2 maps.

Pociask-Karteczka J. (1990): Water circulation in particular seasons in north-west Sørkapp Land. *Zeszyty Naukowe Uniwersytetu Jagiellońskiego, Prace Geograficzne*, 81: 101–125.

Post E., Klein D.R. (1999): Caribou calf production and seasonal range quality during a population decline. *Journal of Wildlife Management*, 63: 335–345.

Prach K., Košnar J., Klimešová J., Hais M. (2010). High Arctic vegetation after 70 years: a repeated analysis from Svalbard. *Polar Biology*, 33(5): 635–639.

Pulina M. (1977): Uwagi o zjawiskach krasowych w południowej części Spitsbergenu. *Kras i speleologia*, 1(X): 104–129.

Salvigsen O., Elgersma A. (1985): Large-scale karst features and open taliks at Vardeborgsletta, outer Isfjorden, Svalbard. *Polar Research*, 3: 145–153.

Salvigsen O., Elgersma A. (1993): Radiocarbon dating of deglaciation and raised beaches in northwestern Sørkapp Land. *Zeszyty Naukowe Uniwersytetu Jagiellońskiego, Prace Geograficzne*, 94: 39–48.

Speed J.D.M., Cooper E.J., Jónsdóttir I.S., van der Wal R., Woodin S.J. (2010): Plant community properties predict vegetation resilience to herbivore disturbance in the Arctic. *Journal of Ecology*, 98: 1002–1013.

Stange R. (2003): Rock and Ice. Landscape of the North. A geographical traveling accompaniment for Spitsbergen and East Greenland (68–74°N). Rolf Stange, 238 pp.

Styszyńska A. (2007): Tabele [in:] A.A. Marsz, A. Styszyńska (eds.): *Klimat rejonu Polskiej Stacji Polarnej w Hornsundzie – stan, zmiany i ich przyczyny.* Akademia Morska w Gdyni, Uniwersytet Śląski, Instytut Geofizyki PAN, Gdynia: 295–331.

Sund M., Eiken T., Hagen J.O., Kääb A. (2009): Svalbard surge dynamics derived from geometric changes. *Annals of Glaciology*, 50(52): 50–60.

Tyler N.J.C., Øritsland N.A. (1989): Why don't Svalbard reindeer migrate? *Ecography*, 12(4): 369–376.

Walker M.D., Wahren C.H., Hollister R.D., Henry G.H.R., Ahlquist L.E., Alatalo J.M., Bret-Harte M.S., Calef M.P., Callaghan T.V., Carroll A.B., Epstein H.E., Jónsdóttir I.S., Klein J.A., Magnusson B., Molau U., Oberbauer S.F., Rewa S.P., Robinson C.H., Shaver G.R., Suding K.N., Thompson C.C., Tolvanen A., Totland Ø., Turner P.L., Tweedie C.E., Webber P.J., Wookey P.A. (2006): Plant community responses to experimental warming across the tundra biome. *Proceedings of the National Academy of Science of the USA*, 103: 1342–1346.

Wassiliew (Vasiliev) A.S. (1925): Spitsberg III. Feuille du Sud. (map) [in:] *Missions scientifiques pour la mesure d'un arc de méridien au Spitsberg.* Miss. Russe, Publication de l'Académie des Sciences de Russie.

Werenskiöld W. (1920): Spitsbergens fysiske geografi. *Naturen* (Bergen), 44: 209–242.

Werenskiöld W. (1953): The Strand Flat of Spitsbergen. *Geografisk Tidsskrift* (København), 52: 302–309.

Winsnes T.S., Birkenmajer K., Dallmann W.K., Hjelle A., Salvigsen O. (1992): Geological Map, Svalbard, 1 : 100000, Sørkapp [in:] *Norsk Polarinstitutt Temakart*, 17.

Ziaja W. (1985): The influence of winds on the temperature and humidity at north-western Sörkappland (Spitsbergen) in summer 1982. *Zeszyty Naukowe Uniwersytetu Jagiellońskiego, Prace Geograficzne*, 63: 107–113.

Ziaja W. (1988): Fizycznogeograficzne zróżnicowanie górskiej części północno-zachodniego Sørkapplandu (Spitsbergen). Doctor's thesis (manuscript), Jagiellonian University, Institute of Geography, 137 pp.

Ziaja W. (1989): Rzeźba Doliny Lisbet i otaczających ją gór (Sörkappland, Spitsbergen). *Zeszyty Naukowe Uniwersytetu Jagiellońskiego, Prace Geograficzne*, 73: 85–97.

Ziaja W. (1991): Fizycznogeograficzne zróżnicowanie górskiej części północno-zachodniego Sörkapplandu (Spitsbergen). Część I: Uroczyska. *Zeszyty Naukowe Uniwersytetu Jagiellońskiego, Prace Geograficzne*, 83: 31–54.

Ziaja W. (1992): Fizycznogeograficzne zróżnicowanie górskiej części północno-zachodniego Sörkapplandu (Spitsbergen). Część II: Tereny. *Zeszyty Naukowe Uniwersytetu Jagiellońskiego, Prace Geograficzne*, 88: 25–38.

Ziaja W. (1999): Rozwój geosystemu Sørkapplandu, Svalbard. *Rozprawy Habilitacyjne UJ*, 343, Wydawnictwo Uniwersytetu Jagiellońkiego, Kraków, 105 pp.
Ziaja W. (2002): Changes in the landscape structure of Sørkapp Land [in:] W. Ziaja, S. Skiba (eds.): *Sørkapp Land landscape structure and functioning*. Wydawnictwo Uniwersytetu Jagiellońskiego, Kraków: 18–50.
Ziaja W. (2004): Spitsbergen Landscape under 20th Century Climate Change: Sørkapp Land. *Ambio*, 33(6): 295–299.
Ziaja W., Ostafin K. (2007): Współczesna przemiana krajobrazu Lodowca Gås i okolicy [in:] R. Przybylak, M. Kejna, A. Araźny, P. Głowacki (eds.): *Abiotyczne środowisko Spitsbergenu w latach 2005–2006 w warunkach globalnego ocieplenia*. Uniwersytet Mikołaja Kopernika, Toruń: 235–245.

List of enclosures (maps)

Technical editor
Jadwiga Makowiec

Proofreader
Jerzy Hrycyk

Typesetter
Hanna Wiechecka

Jagiellonian University Press
Editorial Offices: ul. Michałowskiego 9/2, 31-126 Kraków
Phone +48 12-631-18-81, +48 12-631-18-82, Phone/Fax +48 12-631-18-83